ACCIDENTAL AGENTS

INSURRECTIONS: CRITICAL STUDIES IN RELIGION, POLITICS, AND CULTURE

INSURRECTIONS: CRITICAL STUDIES IN RELIGION, POLITICS, AND CULTURE

SLAVOJ ŽIŽEK, CLAYTON CROCKETT, CRESTON DAVIS, JEFFREY W. ROBBINS, EDITORS

The intersection of religion, politics, and culture is one of the most discussed areas in theory today. It also has the deepest and most wide-ranging impact on the world. Insurrections: Critical Studies in Religion, Politics, and Culture will bring the tools of philosophy and critical theory to the political implications of the religious turn. The series will address a range of religious traditions and political viewpoints in the United States, Europe, and other parts of the world. Without advocating any specific religious or theological stance, the series aims nonetheless to be faithful to the radical emancipatory potential of religion.

For a complete list of books in the series, see page 283.

ACCIDENTAL AGENTS

ECOLOGICAL POLITICS BEYOND THE HUMAN

MARTIN CROWLEY

Columbia University Press *New York*

Columbia University Press
Publishers Since 1893
New York Chichester, West Sussex
cup.columbia.edu

Library of Congress Cataloging-in-Publication Data
Names: Crowley, Martin, author.
Title: Accidental agents : ecological politics beyond the human / Martin Crowley.
Description: New York : Columbia University Press, [2021] | Series: Insurrections: Critical Studies in Religion, Politics, and Culture | Includes index.
Identifiers: LCCN 2021035035 (print) | LCCN 2021035036 (ebook) | ISBN 9780231204026 (Hardback : acid-free paper) | ISBN 9780231204033 (Trade Paperback : acid-free paper) | ISBN 9780231555333 (eBook)
Subjects: LCSH: Environmental management—Political aspects. | Posthumanism. | Nature—Philosophy.
Classification: LCC GE300 .C766 2021 (print) | LCC GE300 (ebook) | DDC 658.4/083—dc23/eng/20211001
LC record available at https://lccn.loc.gov/2021035035
LC ebook record available at https://lccn.loc.gov/2021035036

Columbia University Press books are printed on permanent and durable acid-free paper.
Printed in the United States of America

Cover design: Milenda Nan Ok Lee
Cover image: Eunice merong © Shutterstock

for Nick

for Ella

Ce qui naît dans la naissance n'est pas d'abord le produit ou l'engendré d'un auteur ou de parents, mais précisément, ce qui naît c'est l'être en tant que rien ne le pose et que tout l'expose, l'être toujours singulier.

—Jean-Luc Nancy

CONTENTS

ACKNOWLEDGMENTS

Earlier versions of material included here appeared as "#automaticpolitics," *Diacritics* 47, no. 1 (2019): 136–53 (Copyright © 2019 Cornell University), and "Earthlings Against Latour!," *Parallax* 27, no. 1 (2021): 97–113.

For welcoming this book into their series, and for their constant support throughout its preparation, I offer my warmest thanks to Jeffrey R. Robbins and Clayton Crockett. Wendy Lochner has been the most expert of editors; I am immensely grateful for the skill with which she guided the book to publication. Many thanks, too, to Lowell Frye for invaluable help through the production phase.

While I was completing the revisions to the text, news came of the death of Bernard Stiegler. I had the great good fortune to discuss the ideas explored here with Stiegler on numerous occasions, and—quite apart from his philosophy, without which the argument I build in what follows simply would not exist, as will be apparent—his exceptional generosity and inspirational example were crucial elements in this book's preparation. I offer this work in tribute to his memory.

The development of the thinking I attempt here was decisively supported and inflected by a number of friends and colleagues, with a word or phrase dropped into a conversation, essential reflections and prompts, or unstinting intellectual companionship—and in some cases, with all three. I would like to thank in particular Emily Apter, Andrew Benjamin, Jane Bennett, Jennifer Cobbe, Oliver Davis, Didier Debaise, Michel Feher, Hannah Freed Thall, Karen Gregory, Suzanne Guerlac, Elizabeth Hallam, Tomasz Hollanek, Ian James, Iwona Janicka, Nikolaj Lübecker, Timothy Mathews, Adrian May, Laura McMahon, Giovanni Menegalle, Gerald Moore, Nick Nesbitt, Cillian Ó Fathaigh, Joanna Page (to whom I owe in particular the phrase "reflexivity without transcendence"), Davide Panagia, Mara Polgovsky Ezcurra, Thangam Ravindranathan, Peter Szendy, Ben Turner, Phillip John Usher (to whom—alongside much else—I owe the notion of trees as plastic readers), Anthony Vidler, Luke Warde, Tomas Weber, and Rachel Wooller. Although the relevant protocol obliges me to exempt them all from any significant responsibility for what follows, and however surprised some of them might be to see their name featuring here (although I hope each might recognize something of their contribution), I would like to think of this list as representing part of the alliance thanks to whose work this book has come to exist.

INTRODUCTION

The generalized planetary ecological crisis has not only brought about a significant change in the understanding of human action, it has drastically increased the stakes of this understanding. As the contested concept of the Anthropocene tries to capture the arrival of human activity as a geophysical force, this activity has simultaneously emerged as ever more deeply enmeshed with the existence and actions of all kinds of other beings. While once we might have thought of human action as underpinned by the exceptional capacity of human beings to transcend their environment, now this belief itself appears to have provided a pretext for the instrumentalization of these other beings, their exploitation in ways whose destructive environmental consequences can no longer be ignored. Against this catastrophic horizon, then, how should human action be understood?

Current philosophy and theory offer two answers to this question, frozen in a stand-off between those for and those against the idea of human beings as uniquely endowed with agency, on the basis of conscious intentionality and sovereign will. Some maintain that only this exceptional sovereignty can guarantee effective political action. Others, motivated by a radically egalitarian ontology, reject this exceptionality and offer, instead, accounts of agency as distributed across groupings of human and nonhuman actors. But in their different ways, both positions have run into difficulties. If the first promotes a human exceptionalism whose disastrous consequences are everywhere apparent, the second has repeatedly been accused of lacking a conception of political agency that would be both coherent and effective. On the one hand, more of the same; on the other, at least according to its critics, a worrying lack of traction.

Accidental Agents seeks a way through this impasse. Convinced of the need to retain the core commitment of each side, I argue that this stand-off is unnecessary; that it is possible to understand political agency as both distributed and decisive. Retaining both the radical ontological equality of composite agential groupings and the necessity of prescriptive political intervention and mobilization, in the following discussions I propose an original account of political agency as both irreducibly composite and allowing effective, interruptive action. Developing an account of agency as emerging contingently as distributed across heterogeneous alliances, I argue that these alliances are also the location of decisive action and future-oriented commitment. Building on the considerable achievements of existing accounts of distributed agency, my aim is to contribute to the ongoing, effective engagement of contemporary critical thought with the grave problems of its moment on the basis of the radically egalitarian ontology this moment demands.

DISTRIBUTED AGENCY AND ITS CRITICS

The following pages introduce this study by setting the scene conceptually and contextually and defining the key commitments of the position I will be developing. First, I will sketch the key features for our purposes of theories of distributed agency before introducing the political criticism to which they have been subject and anticipating something of my own position. We will then take a step back, to reflect on concepts of agency in general and political agency in particular, notably in terms of the relation of these to the rich and catastrophic history of human exceptionalism. Then I will return to our present debates via the contextual factors contributing to the recent interest in understanding agency as distributed. In the second half of this introduction, I will set out my own argument in more detail, presenting in particular its terminology, conceptual underpinnings, and key motifs and providing an outline of its structure.

We begin, then, with the salient aspects of recent theories of distributed agency. Broadly speaking, we can characterize these as having come to the fore in the context of the theoretical tendencies loosely identified as new materialist and posthumanist. Michael Haworth offers a good overview of the relation between this intellectual context and the theories in: "In work deriving from such diverse streams of thought as animal studies, systems theory, actor- question network theory, the new ("speculative") materialisms and the varied discourses surrounding cognitive neuroscience, the human is demoted from its privileged position as an ontological exception and situated within a wider ecological network. . . . [Such] fields are concerned to de-emphasize human agency as well as call into question the uniqueness of its form of perceptual access onto the world" (2016:151–52).

With specific reference to the question of agency, the key contribution to the development of this thinking is undoubtedly the work of Bruno Latour, including that undertaken alongside Michel Callon and John Law in the elaboration of actor-network theory. As I will be discussing Latour in detail in chapter 1, I will not do so here. It will be enough for now to note that Latour's adoption of the term *actant*, which he takes from the semiology of A. J. Greimas and uses to suspend the human privilege associated with notions of *actor* and *agent* (see Latour 1988:252n11), signals his commitment to understanding agency not as a given property of a particular set of beings but as an emergent feature of contingent associations—or, in his terms, "alliances"—among all manner of beings (1988:160). This conception has found considerable success in the wake of Latour's work, notably in social theory. Its core tenets are well formulated by Werner Rammert, himself one of the first to use the term *distributed agency*: "actions are composed of many elements, and performing those actions is a process distributed across several acts and actors" and "This collective agency is constituted by the distributed activities of the heterogeneous units" (2012:90, 107).

As Lois McNay helpfully summarizes, then: in this conception, "agency is regarded not as the exclusive property of humans but rather as an ever-changing set of potentialities immanent within the energetic and uncontainable dynamics of material existence" (2016:53). In Anglophone philosophy and critical theory, the most influential concerted development of the notion has come from the work of Jane Bennett, particularly in *Vibrant Matter* (Bennett 2010). As Bennett puts it, in this conception, "the efficacy or effectivity to which [agency] has traditionally referred becomes distributed across [a] heterogeneous field, rather than being a capacity localized in a human body or in a collective produced (only) by human efforts" (2010:23). Bennett's exploration of this model roams compellingly from Spinoza to electrical storms, from Darwin's worms to John Dewey, from Kafka's Odradek to a gunpowder residue sampler; motivated, she says, by a "hunch . . . that the image of dead or thoroughly instrumentalized matter feeds human hubris and our earth-destroying fantasies of conquest and consumption. It does so by preventing us from detecting (seeing, hearing, smelling, tasting, feeling) a fuller range of the nonhuman powers circulating around and within human bodies" (ix).

If Bennett's approach is in part motivated by a concern for greater descriptive accuracy than is permitted by a schema composed only of human actors on the one hand and everything else on the other, it is above all driven by what I have been calling an *ontological egalitarianism*. In Mark Hansen's formulation, this means that "we must rethink agency as the effect of global patterns of activity across scales in networks, where absolutely no privilege is given to any particular individual or node, to any level or degree of complexity" (2015:2). The aim is accordingly that expressed by William E. Connolly: "to appreciate multiple degrees and sites of agency, flowing from simple natural processes, through higher processes, to human beings and collective social assemblages" (2010: 22), where "simple" and "higher" describe degrees of organizational complexity but connote no ontological privilege. On this basis, a continuum of agential configurations comes into view: it becomes possible "to construe human agency as an emergent phenomenon, with some nonhuman processes possessing attributes bearing family resemblances to human agency and with human agency understood by reference to its emergence from non-human processes of proto-agency" (Connolly 2010:23).

This concern for better understanding on the basis of greater descriptive accuracy—"to come to terms more richly with multiple modes and degrees of agency that compose the world," as Connolly puts it (31)—is complemented in Bennett's case by an explicit desire to improve the quality of human action in the world. As she writes, "The hope is that the story will enhance receptivity to the impersonal life that surrounds and infuses us, will generate a more subtle awareness of the complicated web of dissonant connections between bodies, and will enable wiser interventions into that ecology" (2010:4). As a result, Bennett is particularly interested in how such enhanced ecological sensitivity might specifically inform human political decisions. Despite the conventional restriction of political action to more or less autonomous human actors, in her approach, "it is a safe bet to begin with the presumption that the locus of political responsibility is a human-nonhuman assemblage. On close-enough inspection, the productive power that has engendered an effect will turn out to be a confederacy, and the human actants within it will themselves turn out to be confederations of tools, microbes, minerals, sounds, and other 'foreign' materialities. Human intentionality can emerge as agentic only by way of such a distribution" (36). Seeking accordingly to relativize the position of human beings within a broad ecology of "vibrant matter," Bennett hopes thereby "to see how analyses of political events might change if we gave the force of things more due" (2010: viii).

It will already be clear that Latour, Bennett, and Connolly refuse a model of agency as the exclusive property of human beings; indeed, this is the fundamental axiom of their work. For various critics, however, such a refusal renders this work, and other cognate contributions, incapable of the kind of political efficacy we here see Bennett seeking. In a broadly sympathetic account of these contributions, Arjun Appadurai notes, "Most of these thinkers acknowledge that there is some tension between the physics and metaphysics of most variations of this new materialism and our classical ideas of normativity and political critique" (2015:222). Many others are bluntly critical. For Thomas Lemke, "In attacking a humanist account of politics, [Bennett] not only exposes the limits of humanism, but also gets rid of politics" (2018:47). For Adrian Johnston, discussing Connolly,

"The vaguely Heraclitan flux doctrines palpably lurking in the background hardly are conducive to a targeted and disciplined set of political practices" (2014:298). In McNay's verdict, "it is not easy to see how theories of posthuman agency translate into the type of emancipatory and radical political practice that is claimed for them"; such theories, McNay continues, "do not straightforwardly yield a viable account of intervention in the world" (2016:55; and for a good overall account of the political stakes here, see Burns 2016).

The root cause of this supposed political intransitivity is spelled out in extended critiques by Alf Hornborg and Andreas Malm. According to these critiques, the theorists in question would simply have misunderstood the nature of agency. More precisely: in extending the category of agency beyond its proper bounds, they would have confused agency proper with mere causality (specifically, in Aristotelian terms, with efficient causality). For Hornborg, though "objects (and living fetishes) may constrain, prompt, or mediate the agency of living organisms," "in no case is it justified to dissolve the crucial difference between purposive agency and merely having consequences" (2017:99). In Hornborg's model, the division between agency and causality falls between animate and inanimate beings; he happily accepts that inanimate beings influence the animate, but he refuses to consider this influence as a form of agency, however attenuated. And the finality of Hornborg's argument is political: "Only by applying such distinctions," he writes, "are we able to grasp the predicament of the Anthropocene and to expose the exploitative global power relations underlying the ideology of economic growth and technological progress" (96).

Drawing on Hornborg and philosopher Lilian O'Brien, Malm also insists on distinguishing agency from mere causality: "The meteorite makes some difference to a state of affairs, but that is a definition of *causal impact*—not agency, which is a subclass of things that make a difference" (2018:96). For Malm (here departing from Hornborg), this subclass is defined quite classically by "having a mind" (85); specifically, the capacity for futural projection: beings with minds act, beings without minds produce effects. And what matters to Malm is, of course, that only the strongest, most purposive kind of action, on the part of the beings who alone have minds, has a chance of resisting, let alone reversing, the ongoing human exacerbation of the climate emergency. "A resistance," he writes, "can be conceived only by affirmation of the most singularly human forms of agency" (108).

For the most part, I consider these accusations ill-founded—for two reasons. First, most of those they target (notably, here, Bennett and Connolly) are not proposing a general model of political agency but are inviting human political actors to understand and situate their own interventions more attentively as part of a broad field shaped by the contributions of all kinds of actors and factors. In Bennett's words: "The task at hand for humans is to find a more horizontal representation of the relation between human and nonhuman actants in order to be more faithful to the style of action pursued by each" (2010:98). Or, in the mode of the categorical imperative, "Seek instead to engage more civilly, strategically, and subtly with the nonhumans in the assemblages in which you, too, participate" (116). Or, as Connolly puts it, "To appreciate human entanglements with a variety of nonhuman forces . . . may help to ennoble the larger ethos in which we participate" (2017:61). As Diana Coole writes, then, despite the fears of the human exceptionalists, this approach "does not preclude an identification of agents who might manifest their capacities in ways which have a strong affinity with conventional accounts. It is merely that their emergence has to be traced and not presumed, which will likely result in their capabilities for agency being recognised as more partial, contextual and provisional than liberal humanism (individual agency), Marxism (class agency) or realist approaches to International Relations (state agency) allow" (2013:457–58).

Indeed, to accuse exponents of this approach of abandoning the specificity of human agency is to fail to read them. Both Bennett and Connolly, for example, explicitly grant a limited privilege to human concerns within their approaches, as an acknowledgment of the situated inflection of existential priorities. Bennett's analyses are "motivated by a self-interested . . . concern for *human* survival and happiness." "I cannot envision," she writes, "any polity so egalitarian that important human needs, such as health or survival, would not take priority" (2010: ix–x, 104). Accordingly, though Bennett, drawing here on Latour, "strategically elides what is commonly taken as distinctive or even unique about humans," she explicitly does so "for a while and up to a point" (ix). For his part, Connolly remains wary "of any version of 'posthumanism' susceptible to the charge that it does not give *any* significant priority to the human estate in its multiple entanglements with other beings and processes" (2013:13).

If they certainly refuse the crude division between mere causality and agency proper, then, these theorists openly distinguish human action from that of other actants. For Connolly, as we have seen, this distinction concerns degrees of complexity: other actants should be understood as "micro-agents" (2013:85) or "proto-agents" (2010:24), or as displaying "minimal agency" (2010:26), whereas "complex agency" remains reserved for humans (2010:26). In this view, that simpler levels are nested within more complex levels means both that the more complex must be understood as constitutively entangled with the more simple and that the two remain significantly distinct. The exemplary attention Bennett and Connolly pay to the details of this entanglement should not confuse us when it comes to the finality of this approach: the goal remains wiser, better informed human action. In Bennett's eloquent declaration, "Agency is, I believe, distributed across a mosaic, but it is also possible to say something about the kind of striving that may be exercised by a human within the assemblage" (2010:37–38).

Their human-exceptionalist critics thus misunderstand theorists such as Bennett and Connolly in two crucial ways. These theorists certainly do not abandon any meaningful distinction between human and nonhuman forms of agency; and they are proposing not a general theory of political agency in terms of distributed agency but a reassessment of human action on the basis of its richer embedding in a more graduated agential field. This in turn leads me to my second disagreement with these critics: for even when the goal *is* a general theory of political agency, I am not convinced that an understanding of agency as distributed is necessarily incompatible with an account of effective, even antagonistic political agency. It is true that most of those who take this position hesitate to think of politics in these terms, and understandably so: as Connolly writes, "enhanced sensitivity to what is most fragile about ourselves and our place on the planet does not go smoothly with militancy" (2013:10). But Connolly introduces this acknowledgment precisely because he does connect his insistence on broad ecological entanglement to the need for militant action; and his accounts of the forms such action might take—encompassing "creative citizen movements, enlarged state action, interstate agreements, and global citizen actions," up to and including "cross-regional general strikes" (2013:67–68; 2017:129)—are invariably detailed and reasoned. To this extent, then, we must again recognize that the human-exceptionalist critics have overstated their case.

Two qualifications are nevertheless in order here, which give this case what I believe to be its only valid element. First, as per his persistent distinction of human agency from other forms, Connolly's militant actions are specifically those of human activists: he is calling for militancy as part of that more attentive human action we have just seen to be the finality of his arguments. (As he says, "This care for being can be situationally joined to political militancy, if and when events threaten the integrity of that which you care about the most. And that militancy will also be inflected by the underlying sensibility infusing it" [2013:124–25].) In this sense, then, his example does not yet prove that a general theory of political agency as the conjoint agency of diverse participants is indeed compatible with strong, even antagonistic political effectivity. Second, in the case of the only theorist of distributed agency who is actually aiming to describe political agency in terms of this theory—namely, Latour—we do in fact find a significant difficulty when it comes to integrating decisive effectivity and combative intervention into this description. And I will argue in chapter 1 that this significantly weakens Latour's attempt to provide an account of distributed political agency—indeed, that it leaves him unable to conceive of the confrontational action he knows is currently necessary.

Taking these two points together, the human-exceptionalist critiques of theories of distributed agency might be justified in the following sense only: if the aim is indeed an account of politics qua politics, such an account will have to include the dimension of antagonistic conflict. My contention, however, is that this can and must be provided from within a distributed agency approach; that is, without appeal to the supposedly exceptional capacities of ontologically transcendent human beings.

The approach I develop here differs from those of Bennett and Connolly, then, in that it seeks to provide an overall account of political agency in terms of distributed agency. I will seek to suspend any a priori typology of the contributions of this or that kind of being, even in the supremely nuanced and attentive form this takes in their discussions, and instead to map the emergence of political agency through the conjoint contributions of various participants. (I will return to this distinction in horizon 1, in particular the difference between my approach and Connolly's advocacy of militancy.) Though I appreciate their emphasis on irreducible existential situatedness and stand in awe of their patient and fine-grained accounts, I remain nervous about any axiomatic claim for human distinction, however nuanced and entangled. Only with the greatest difficulty, I think, is it possible in this area to prevent this kind of distinction (in the sense of careful differentiation) from drifting toward distinction as elevation, as categorical species privilege and ontological transcendence. As Bennett writes, we face here "the difficulty of theorizing agency apart from the belief that humans are *special* in the sense of existing, at least in part, outside of the order of material nature" (2010:36–37).

As it happens, I consider that the work of both Bennett and Connolly tackles this difficulty more successfully than that of any other Anglophone theorist; and I certainly do not believe that they reintroduce anything like human ontological transcendence. (For Connolly's rebuttal of the charge that his gradations of agency reinstall an ontological hierarchy, see Connolly 2010:31.) Nevertheless, my attempt here will be to build a model that gives not even the slightest houseroom to a priori human distinction on the basis of given attributes, however these are defined. (For this reason, incidentally, I will be somewhat promiscuous in my use of *actor* and *agent*, because, unlike many—not least Connolly, Coole, and Katherine Hayles [2017]—I am not looking to preserve a definitional separation between "complex" agency and other forms.) The challenge, then, will be to build such a model while managing to account for decisively effective political agency. As I have suggested, and as I will show in detail in chapter 1, in my reading, Latour manages the first part of this but not the second, which is where my approach also differs from his, in that it seeks to integrate the possibility of conflictual agency into its account. The difference between my position and our existing theories of distributed agency, then, lies in my attempt to understand specifically political agency both without reference to any a priori human distinction and in a way that is compatible with decisive, even confrontational intervention. I hope it will be understood that I am seeking not to supplant these existing approaches but to supplement them, extending their insights to a dimension of the problem that in my view requires further attention.

Overall, then, I am looking to build a model of distributed political agency in which decisive, confrontational action will certainly not be the whole story but in which such action can find a coherent place without being restricted to human activism alone. I will set out the core elements of this model, including my reasons for wanting to maintain the possibility of conflict as part of its scope. First, however, we should spend a little time considering the concept of agency itself and the prehistory of these recent debates.

AGENCY, POLITICS, AND HUMAN EXCEPTIONALISM

From Aristotle's *Nicomachean Ethics*, to Kant's *Groundwork of the Metaphysics of Morals*, to contemporary moral philosophy, the Western tradition has consistently defined meaningful agency in terms of autonomy (itself understood through the twin dimensions of informed understanding and freedom from constraint), intentionality, and rational choice. (For an excellent, succinct account of this, see McNay 2016.) Though the precise definitions of these core concepts might vary, their combination is consistently grounded in an understanding of agents as in a strong sense purposive: namely, equipped with reflexive consciousness (the "I think" that, for Kant, "must be able to accompany all my representations" [1998:246]) and able to interpret their surroundings, project their intentions into the future, make decisions based on these projections, and take responsibility for the consequences of these decisions. And this "capacity concept of agency" (Passoth, Peuker, and Schillmeier 2012:1) invariably in turn accompanies a human-exceptionalist claim. As Balibar and Laugier write, "agency is supposed to be what characterizes, among the events of the world, what belongs to the order of human action" (2014:19), distinguished by what Brian Massumi eloquently calls "those capacities over which we human animals assert a monopoly, and on which we hang our inordinate pride in our species being" (2014:2).

From thermostats to whales, various other beings may exhibit lesser versions of the requisite capacities, but only in human beings do we find the full array developed in full (see O'Brien 2015:136–45). Crucially, this plenitude comes to constitute a difference not only of degree but of kind: a step change thanks to which human beings are fundamentally distinguished from all others and defined as properly transcendent in relation to their environment. Examples of this conception abound: we might think of Francis Bacon's project to reestablish the human "empire over creation," lost in the Fall, via correct knowledge of nature and the command this will permit (1902:290); Descartes's subsequent, oft-cited promise that such knowledge "of all other

Bodies which environ us" will allow human beings to "employ them to all uses to which they are fit, and so become masters and possessors of Nature" (1648: part. VI, para. 2); and, in its paradigmatic form, the consequent assertion shared by Kant and Diderot that thanks to their superior abilities—above all the capacity for reflexive thought—humans are separated from other animals by an *infinite* distance (Kant 2012: 15; Diderot 1984b:228). As McNay puts it, "the mode of agency as transcendence" effects "a break with the world that institutes a new kind of being" (2016:41): that split by virtue of which humans claim for themselves the freedom that entitles them to either exploit or take care of all other beings (and generally do both), these other beings remaining, in either case, law-bound creatures of necessity—"with which," as Kant puts it, "one can do as one likes" (2012:15).

This is the model we in the West inherit from our traditional substance metaphysics, with its categorical attribution of certain qualities to certain types of beings, which works to secure human beings as ontologically transcendent by defining their capacities as exceptional. (Even if this exceptionality is formulated as a unique deficiency in need of repair by such capacities, as in the myth of Prometheus, or a unique incompletion that such capacities come to make good, as in subsequent notions of human perfectibility.) Within, alongside, and beyond this tradition, there are, to be sure, theological and philosophical positions that seek to rethink transcendence as a function or even a dimension of radical immanence, from—to take a small handful of indicative names—Nicholas of Cusa to Baruch Spinoza to Jean-Luc Nancy. And such positions do, moreover, tend to prove compatible with a more egalitarian model of the relation between different forms of existence. Inasmuch as these positions remain attached to an a priori distribution of capacities, however, this egalitarianism must remain limited. The post-Heideggerian tradition (within which we might situate Nancy) is symptomatic here: its displacement of glorious human ontological distinction, and its accompanying enfolding of transcendence as the immanent opening out of the world, are arguably vitiated by a persistent appeal to uniquely human capacities, through which this opening out is exclusively articulated. (We might think here of Heidegger's restriction of the encounter with beings "as such" to that *Dasein* which, for all that it might not be classically anthropological, is never other than human; or of Nancy's residual retention of human language as the privileged locus of the exposure of the sense of the world; on these, see Crowley 2019.) So although the question of transcendence can, of course, be modulated in various ways, and claims to radical or pure immanence remain core features of contemporary philosophy and theology (see in particular Laruelle 2013; and Hallward 2006), the form of transcendence that sets the stakes of this study remains the strong ontological transcendence invariably generated by a substance metaphysics of a priori capacities and the split this imposes between human actors and all other kinds of beings.

In the case of specifically political agency, this transcendent split becomes particularly acute, in two senses. First, however *politics* is defined, it is overwhelmingly agreed to be an exclusively human activity. (One proof of this *a contrario* is the defamiliarizing effect of a title such as Massumi's *What Animals Teach Us About Politics* [2014].) Insects can be gregarious, dolphins can be social, but there is only one political animal. Indeed, political activity (again, however this is defined) is invariably one of the preeminent attributes—along with the consubstantial trinity of reason, morality, and language—by which human beings like to distinguish themselves infinitely from other kinds of beings. Diderot is again here paradigmatic: in his account of natural law, having played with the long-standing allegorical image of a political assembly of nonhuman animals, he uses the evident nonsense of such a scenario (when taken literally) to underscore the inviolable boundary between humans and these others, and he affirms politics as proper only to human beings, the product and the foundation of their unique dignity (1984a:146).

As Latour puts it, ventriloquizing and historicizing this position, "Obviously there is no politics other than that of humans, and for their benefit! This has never been in question. The question has always been about the *form* and the *composition of this human*" (2018:85). Second (and consequently), the activities undertaken under the name of politics are themselves frequently defined in terms of those forms of cognitive or temporal transcendence to which human beings lay exclusive claim. For politics, we might say with Badiou, is a matter not just of decisions on how to live together (a definition, we should note, that already presupposes reflexive consciousness and the projection of intentions); politics is a matter of commitment and prescription. And for those—like Badiou—who are happy to sign up to the human-exceptionalist position, the self-awareness and futural projection these require (in Badiou's blunt terms: the capacity for thought) stake out politics as always and only a human activity (see Badiou 2005b:97–98). What is more, political modalities such as commitment and prescription are not only habitually secured by the exclusively human capacity for one form or another of transcendence but can often themselves be defined in terms evocative of transcendence, as forms of rupture or radical break. Badiou is again the best example here: "The essence of politics is not the plurality of opinions," he writes: "It is the prescription of a possibility in rupture with what exists" (2005b:24).

This is not the place to go into the extensive arguments over whether or not the Event, from which this militant prescription follows as an act of fidelity, is in Badiou's account transcendent in relation to its situation. (See Badiou 2005a; and, for a good account of these arguments, Gibson 2012:256.) For our purposes, what matters is that models of the political event as "a pure cut . . ., an atemporal instant that separates a past from a future and extracts a time from another (indifferent) one" (Gibson 2012:45), can seem to advocate a political or temporal transcendence complicit with the abusive fantasies of human ontological transcendence. (On the qualities and shortcomings of such models, see especially Apter 2018.)

As we will see in detail with reference to Latour (in chapter 1), this potential complicity can in turn lead thinkers motivated by a concern for ontological egalitarianism to reject models of politics that include strongly conflictual intervention. "Sheathe your swords!," they say. "No more clear-cut bloody decisions" (Callon, Lascoumes, and Barthe 2009:11) Despite the obvious appeal of such a refusal of political violence, however, with its recommendation of careful, piecemeal negotiation and its ironic dismissal of "Manly warrior assurance" (Callon et al. 2009:11), it is not certain that frontal conflict can indeed be excluded from the field of politics. As I will argue later, we do not have to be normatively committed to a politics of trenchant effectivity to find the refusal of confrontational intervention a flaw in what wants to be a general theory of politics. Even if all we want from a model of political action is descriptive accuracy, a model that rules out such intervention as a matter of principle is bound to come up short. In its starkest form, then, the question that motivates my considerations here is: can we coherently understand decisive political intervention from within a fully distributed conception of agency? And the wager of this book is that we can answer this question in the affirmative: that a politics without transcendence is indeed possible.

HOW DID WE GET HERE?

Shortly, I will introduce the core elements of my argument. First, however, some more context is needed to situate our present debates against their broader backdrop. If the idea of dispensing with both the human-exceptionalist claim to sole possession of strong agency and the related fantasy of ontological transcendence that have recently gained significantly in prominence, this is not just thanks to signal contributions such as those of Latour and Bennett; it is also as a result of the interactions between such specific intellectual contributions and broader contextual factors. The situation is well described by Didier Debaise: "Does our contemporary experience not force us to quit a purely anthropological paradigm in order to elicit the centers of experience, manners of being, multiple relations that existents have with each other, and which make up a nature that has become essentially plural?" (2017:41).

If Debaise's reference to nature seems to foreground the ecological dimension I have already invoked, its scope is in fact larger (in keeping with the use of this term in Debaise's key framework, namely the process metaphysics of Alfred North Whitehead). Accordingly, other defining features of "contemporary experience" join with this ecological dimension to form the sense that such an exit is necessary. And given the importance of the theater of mental representation to the "anthropological paradigm," it is no surprise that the most prominent among these features are those theoretical, technological, and experimental developments that have put extreme pressure on the exceptionality this theater was designed to stage.

Alongside the dramatic recalibration of the relation between human action and planetary forces promoted by the concept of the Anthropocene, then, the impetus to understand agency as distributed has been created in particular by the arrival of cybernetics as a dominant epistemology and, in obvious tandem with this, advances in artificial intelligence, especially in the field of deep learning (i.e., that part of research into machine learning in which artificial neural networks engage in relatively unsupervised learning). (As I will discuss artificial intelligence more fully in horizon 2, I will not go into much detail here.) According to Rammert, the concept of distributed agency can be linked in particular to "the many levels and parallel processes of 'distributed computing' " and "the self-organized adaptation processes of 'distributed cognition' " (2012:90). If we might link this latter field to the still-anthropocentric notion of the "extended mind" (see Clark and Chalmers 1998), or arguments for understanding "thought" as a capacity of nonhuman beings (as in *How Forests Think* [Kohn 2013], for example, or *Plant-Thinking* [Marder 2013], or even *Cognitive Biology* [Auletta 2011]), its more significant context is the interest in military and other research in "swarms, distributed intelligence," and especially "insect models of organization" (Parikka 2010: xi) and the "extended organism" (Srinivasan 2018: n.p.), whose inductive epistemology is often better able to resist a priori substance-metaphysical categories.

As the exorbitant expansiveness and, in part, the deadly aims of such developments suggest, the importance of cybernetics as a factor in the challenge to the human-exceptionalist conception of agency can hardly be overstated. As information technology established itself as central to modern human existence, the cybernetic treatment of meaning as information, and of understanding as data processing, served as its accompanying epistemology. (On the significance of cybernetics as epistemology, see Pickering 2010; and Rid 2016.) Most important, this shift moves the emphasis in the modeling of action away from intentionality (whose constitutive framework of mental representation supports the primacy of meaning and understanding) toward a post hoc functionalist description of the operative elements defining a given system. This is well captured by Gregory Bateson, whose participation in the earliest theorization of cybernetics at the Macy Conferences (1946–53) contributed to his development of a pioneering model of agency as distributed across a heterogeneous coalition of contributors. In Bateson's words, with this shift from meaning to information, "thinking, acting, and deciding" are now understood as located in the "cybernetic system—the relevant total information-processing, trial-and-error completing unit" (1972:331, 467). Bateson in fact builds this out to a thoroughgoing tripartite *ecology* of human individual, society, and ecosystem, in which action issues from the relevant system—and even in the case of human individuals, "this 'system' will usually *not* have the same limits as the 'self'" (317; Bateson's example here is a woodcutter felling a tree with an axe).

From Bateson's cybernetic ecology, we need now to bring in the life sciences proper in order first to note the belated influence of Jakob von Uexküll's "phenomenological biology" (Smith 2013:1). For Uexküll, every organism actively processes the meaning of those elements that appear to it as significant in the relay between its "perceptor world" and its "effector world" (Uexküll 1957:6). His proto-cybernetic model later receives detailed systems-theoretical elaboration in the "biology of cognition" of Humberto Maturana, in which the relation of structural coupling between an organism and its environment is allied to the self-conscious inclusion of the observer in the system observed, in the classic manner of second-order cybernetics, to develop an account of self-fashioning (or *autopoiesis*) in which autonomy is rethought as immanent to processes of emergence—and in which, crucially, it becomes possible to think reflexivity without transcendence (see Maturana and Varela 1980).

The emphasis in second-order cybernetics on questions of reflexivity and self-fashioning, especially as extended to the realm of theoretical biology, indicates the scale of its challenge to the thesis of human exceptionality. Inasmuch as this thesis is grounded above all in the exclusive human capacity for reflexive thought, and the cognitive and moral autonomy this is imagined to guarantee, the theorization of reflexivity outside of the theater of mental representation deals it a severe blow. If we now combine the profound implication of human beings in their planetary milieu brought about by the realization of the climate emergency with the fundamental epistemological displacement effected in this way by cybernetics, and the prominence of related developments in artificial intelligence, we get a sense of the extent of the broad systems-theoretical recontextualization of human action over the past half century or so. Against this backdrop, it is perhaps easy to see why the idea of agency as distributed across diverse networks of beings of all kinds has been able to gain such significant traction.

To switch briefly from this broader cultural history to a narrower version of intellectual history, I close this section with a word about a key figure in the immediate prehistory of the current debates that form the context for this study's contribution: namely, Michel Serres. In composing the book's corpus, I made the decision not to feature Serres, despite his significant status as a major and, until lately, underacknowledged forerunner of this field. (For an excellent overview of Serres's considerable *oeuvre* and, in particular, reflection on the reasons for his relative lack of prominence in the Anglophone world, see Watkin 2020, especially 12–18.) Although his work certainly did anticipate and, to an extent, influence more recent interest in thinking agency across networks of human and nonhuman actors (especially in the case of Latour, and not least in its engagement with the implications of cybernetics), the nature of this work, in which different concerns are often woven into rich and densely allusive skeins, means that this question is not thematized in these writings with the same direct, precise focus as Latour brings to bear. We could even say that Serres is not much given to thinking agency as a specific, discrete question—although he is certainly concerned to think expansively about the relation between humans and other kinds of beings, and to think ambitiously about the kinds of framing these relations deserve. (For an authoritative account of Serres's approach to these questions, see Watkin 2020:341–50. I note in passing that, suggestively, the index to Watkin's superb study contains no entry for agency and none for politics.) More particularly, the core concern of this study—namely, the politics of distributed agency—has established itself as a key contemporary problematic in and around the work of Latour, along with that of Bennett and Connolly. Given that this study is designed as an intervention into these contemporary discussions, I have accordingly taken the decision to frame it in terms of these existing points of reference rather than expanding these to include Serres as one of their major antecedents.

TERMINOLOGY

In briefly sketching these contextual factors, however, we have done more than just identify the background to this success. By invoking both the planetary climate emergency and the hegemonic power of digital computing, we have also identified the most significant dimensions of the geopolitical framework that sets the stakes for any current understanding of political agency. As I have said, my aim here is to formulate an account of distributed political agency that contributes to the engagement of contemporary critical thought with this framework by its distinctive double gesture: on the one hand, honoring the ontological egalitarianism demanded by this context through refusing any substance-metaphysical reliance on the predetermined capacities of given types of beings, and on the other, understanding distributed political agency as decisively effective. The time has now come for me to set out the commitments that will underpin this account and to give a sense of what it will look like.

In broad terms, the model I will develop understands agency in the following terms (I emphasize here the elements I will later gloss further): as an *emergent* feature of an *antagonistic alliance* of diverse *participants* within the *array* of beings making up a given situation. Further, to allow this to function as a model of specifically political agency, I will understand such alliances as the site of a *decisive* intervention. I use "decisive" to evoke two principal dimensions of such an intervention: first, that it makes a major difference to a situation (as in, "a decisive contribution"), in some cases forming a confrontational alignment; and, second, that it effects a decision in the strong sense (as in, "they acted decisively"), taking a position and engaging a partisan futural commitment. A decision, that is, just as understood by Badiou: as opening a bifurcation, adopting one side of an either/or alternative (see Badiou 2009:399–447; and on this, Galloway 2020). These two senses will be established, in this order, principally in the book's two horizon sections: first, a largely post hoc, descriptive dimension, seeking to capture the emergence of agency as processually immanent, significantly effective, and confrontationally aligned; second, a more futural dimension, configuring this emergence as entailing a commitment. And, of course, it will be fundamental to my argument that *decisive* action (in both senses) is the action of a composite alliance of various participants.

In the model I will develop, then, actions are effected by alliances. (I will preserve this Latourian term, albeit for reasons that pull away from his approach, as I will explain.) Alliances are composed of some of the diverse array of beings that make up a given situation. For any given action, agency is accordingly a "property" of the relevant alliance. It is therefore not a matter of identifying within the alliance which entity has this or that degree of agency. (Which is why I will be relatively promiscuous in my use of "actor" and "agent," as mentioned earlier.) That being said, different participants will have contributed in different ways and to different extents. The future perfect is here significant: participants become participants, as opposed to the beings making up a situation, only *in* the relations that shape the action in question. In Verbeek's helpful summary of Latourian emergence, "Actants must not be conceived as free-standing entities that then enter into relations with each other. Only in these relations do they become actants; they 'emerge' within the networks that exist between them" (2005:149). What is more, as I will argue in horizon 1, the particular capacities of any participant are similarly established by their differential distribution throughout the alliance. Identifying agency is accordingly a matter of configuring the differential distribution of activities and capacities across the alliance, mapping the eminent sites of accountability (see Floridi and Sanders 2004) that emerge through its action.

I use *array* and *alliance* in place of the more familiar term *assemblage*, as used especially by Latour and Bennett and derived from Deleuze and Guattari's concept of *agencement* (see Deleuze and Guattari 2014). In the uptake of this figure in models of distributed agency, the assemblage appears as "an endless, nonhierarchical array of shifting associations of varying degrees of durability" (Appadurai 2015:221). A given element might participate in this or that assemblage, and although it is the assemblage that is the operative unit, this is thanks to the contingent relations among its elements that here acquire stability but will subsequently be reconfigured. As Hayles puts it when defining her use of the term, "the configurations in which systems operate are always in transition, constantly adding and dropping components and rearranging connections" (2017:2). Relations are exterior to their terms; assemblages are temporary, aggregate wholes "whose properties emerge from the interactions between parts" (DeLanda 2006:5).

If this uptake of the concept has been severely criticized as a fundamental misreading of its Deleuzo-Guattarian usage (see Buchanan 2015), the severity of this criticism explicitly acknowledges the extent of the appropriation in question: in social theory in particular, assemblage theory has become a major paradigm. My reasons for not adopting it here are related to this criticism, however, inasmuch as my argument requires terminology that will allow the recognition of differentiated locations of effectivity, which an emphasis on transient associations can dissolve into equivalence. Contrasting the array to the alliance thus permits a distinction between the beings that happen to make up this or that situation and the subset of these beings that emerges as agential through the action in question. In one sense, then, we might say that array, with its sense of contingency, corresponds quite well to "assemblage." Alliance, on the other hand, brings with it a greater sense that the coming together of these beings has proved in some way decisive, not least in that these beings have now become *participants*. That is, the emergence of an alliance from within an array brings forward what Alfred North Whitehead calls "a novel togetherness" (1978:21): it is not just a loose association of autonomous elements that remain unchanged by their coming together.

In this way, my use of "alliance" differs from Latour's. If Verbeek is right that Latourian actants emerge only in their constitutive networks, it is also the case that Latour can nevertheless succumb to an atomistic version of the alliance, in which its elements can precede and be left intact by its composition and which, for this reason, looks much more like an assemblage. (For a critique of Latour in these terms, see Ingold 2016:13.) Conversely, I will describe the agency of my alliances as "distributed" rather than Latour's habitual term, "hybrid." Partly this is because of the term's implication in racist colonial discussions of so-called miscegenation (see Elam 1999)—although I emphatically do not consider Latour in some way guilty by association as a result of this. Mostly I use "distributed" to underscore the differential distribution of capacities in the composite action of the alliance, which the suggestion of amalgamation in "hybrid" can obscure. But most significantly, I depart from Latour's conception in my insistence on the alliance as—in some cases—a specifically *antagonistic* formation; and this conflictual dimension can again be lost when we think in terms of hybridity. I will be arguing that Latour's inclusion of conflict on his initial spectrum of modes of negotiation soon gives way to an inability to address the reality of antagonism, which his writings invariably elide. Against this, I will develop an account of alliances arraigned in conflict, through which it becomes possible to identify the emergent agency of the alliance as compatible with effective, confrontational political agency.

This emphasis on antagonistic agency brings me to a fundamental question raised by my broad sympathy for the ontological egalitarians: why keep the category of agency at all? Given that this has historically been one of the qualities alleged to elevate humans above all other beings, and that it has consequently been a significant contributor to the manifold violence this elevation has permitted, would not any commitment to ontological egalitarianism do better simply to jettison it, along with the rest of the human-exceptionalist machinery? And if this is true of agency, then how much more so of specifically political agency, where that machinery is in overdrive! In expanding the notion of meaningful agency beyond human beings alone, are we not indulging in what Claire Colebrook (2014) calls "ultra-humanism," generously granting our coveted human qualities to other beings, still considering these only inasmuch as they resemble us in some way? Or flattering ourselves that we can bring to life otherwise inert matter by adding a sprinkling of our magical human dust, in Ingold's marvelous image (2011:29)? When we think we are at our most inclusive, is it not then that we are at our most hubristically human-exceptionalist? (On this, see also Kirby 2016.)

These points are well made and well taken. An approach that works from a schema of human and nonhuman actors, with the goal of including the latter in activities once reserved for the former, will indeed struggle to escape the pull of human exceptionalism. For as long as we are discussing the qualities or capacities of determinate kinds of beings as if these were categorically given, we will remain within this gravitational field, condemned—as Nietzsche pointed out, and as Massumi brilliantly reemphasizes—to cleave the actor from the act and to shrink the latter to a mere manifestation of the former's transcendent capabilities. (See Nietzsche 1998:19; and Massumi 2014:41–42.)

It is for this reason that, as I will shortly discuss, the broad metaphysical commitment underpinning my argument here rejects this kind of substance metaphysics in favor of a metaphysics of process and emergence. I will not, then, be working with a model of distributed agency in which we would start with a property (agency) belonging to a determinate kind of being (humans), which we would then distribute in more or less dilute form to others (nonhumans), like alms to the poor. (Or, in this case, to the "poor-in-world," in Heidegger's term for nonhuman animals [1995:176–273]—which is itself a classic example of such graciously diluted distribution.) But if I nonetheless want to persist with the concept of agency, it is precisely because I want to keep in play all the substance-metaphysical contraband this concept smuggles in. I want us to be thinking in terms of decisions, responsibility, commitment, and so on, because, as I will argue, I am convinced that an account of politics qua politics that abandons these is both descriptively incomplete and, even on its own terms, strategically ineffective. What I will claim, then, is that we can indeed keep this armature of decisive effectivity even as we cut the ties binding agency to the human-exceptionalist theater of intentionality.

The move outside of this theater often provokes a kind of substance-metaphysical panic (some instances of which we have already seen): if strong agency is no longer indexed to human beings, does this not abandon all differentiation in favor of some Hegelian night in which all agential cows are black? Or, more to the point, in which cows (of any color), vacuum cleaners, human beings, and whirlwinds all get to count equally as agents? And does this not then make a mockery of any idea of politics? (Whence Diderot's presentation of a literal parliament of nonhuman animals as evident nonsense, as we saw earlier.) Against such panic, my own approach will, on the contrary, insist on agential differentiation—indeed, on *more and better* differentiation than is offered by a model that knows only two terms: human actors and everything else. The problem is not distinction, not even distinction between capacities. Faithful here to Latourian actor-network theory, I hold that when it comes to the differentiation of forms, modes, or degrees of agency, the problem is, rather, the attribution of certain capacities—more or less attenuated—to certain kinds of beings, in advance of and without regard to any particular situation.

Debaise and Stengers put this very well with reference to Whitehead: such a position, they write, "affirming that nothing must be excluded, does not for all that state that everything must be taken into consideration: it stipulates that we must reject the right to disqualify" (2017:15). I situate the beings that compose my arrays along an unbroken ontological continuum, and I agree with Hayles that "agencies exist all along this continuum, but the capacities and potentials of those agencies are not all the same and should not be treated as if they were interchangeable and equivalent" (2017:67). In my account, moreover, these capacities and potentials are not given but are themselves differentially distributed functions of the relative positions of participants in their respective alliances. Not less differentiation, then, but *more and better*, with a sharper analytical edge. As Karen Barad writes, this position "means that accountability requires that much more attentiveness to existing power asymmetries" (1997:219). This reference to power might surprise, in the vicinity of a theory of distributed agency; as their various critics suggest, do these theories not abandon such trappings of a merely human politics? But how to determine power, if not as a differential? And how to ascertain a differential, if not by more and better, or what Ingold calls "interstitial," differentiation (2016:13)? To quote Latour, this approach "aims not only at establishing equality . . . but at registering differences . . . and at understanding the practical means that allow some collectives to dominate others" (1993:107–8).

CONCEPTUAL UNDERPINNINGS AND KEY MOTIFS

A particularly charged version of such differentiation for this project entails the fundamental need to account for the appearance of agency from within a situation, of an alliance from among an array, in the absence of those human-exceptionalist mechanics of intentionality. That is, I have to respond to the challenge formulated by Hannes Bergthaller, when he writes, "The new materialists . . . have pushed into the background the problem of how sharp ontological and ethical distinctions can emerge *immanently*, as a result of material self-organization" (2014:40). Granted, in some ways this is not my problem: I am not locating this study under the banner of the *new materialisms* (nor, indeed, of posthumanism), and I am certainly not understanding agency as a sharp ontological distinction. But I do need a way of conceptualizing the arrival within a situation of a feature (here, the agency of the antagonistic alliance) not previously given by that situation. As my vocabulary so far might have indicated, I find this in a combination of *process* and *emergence*.

For the reasons I have explained, the starting point of the inquiry I pursue in this study involves rejecting a metaphysics of substance, with its belief in atomized individual things whose properties are given in advance by the category to which they belong. As may fairly be concluded, my thinking in what follows is broadly underpinned by a metaphysics of process, which considers such things not as primary givens but as "manifolds of process" (Rescher 1996:51). Such an approach seeks to account for the existence of individual beings by understanding the process of their composition, and it is motivated by a conviction that heuristically, this process has primacy over the resulting individual. It consequently regards the substance-metaphysical preoccupation with ranking the capacities of given types of individual as unjustifiable. Its claim is that despite the considerable and real differences between actually existing individuals, the fundamental dynamic of their appearance is the same, with the result that all exist on an unbroken ontological continuum. Whitehead, the great pioneer of this approach, makes the point with characteristic elegance: "They differ among themselves: God is an actual entity, and so is the most trivial puff of existence in far-off empty space. But, though there are gradations of importance, and diversities of function, yet in the principles which actuality exemplifies all are on the same level" (1978:18).

For my purposes here, then, the great advantage of this approach is plainly that it allows us to suspend a priori commitments as to what kind of being can exhibit what kind of property, escape the gravitational pull of human exceptionalism, and replace substantial individual things as anchors of agency with composite alliances that emerge during the actions in question as "processual complexes possessing a functional unity" (Rescher 1996:53). And, crucially, as Whitehead's careful reference to gradation and diversity insists, it allows us to do this while maintaining a clear sense of the differential distribution of effective capacities across the participants in these complexes.

Once we have set aside the convenient hierarchies of substance metaphysics, however, we are faced with a further question: how are we to understand the arrival of such differentiation? That is, how are we to answer Bergthaller's challenge and account for the transformation of beings (milling about in an inconsequential array) into participants in an antagonistic alliance, and the constitution of this alliance as an internally differentiated locus of agency? The answer is the concept of emergence—specifically, that of "strong" emergence. In scientific and systems-theoretical discussions, this concept is less common than its twin, "weak" emergence, which is both intuitively less forbidding and empirically more observable; more speculatively compelling, strong emergence is, however, more prominent in philosophical discussions (Chalmers 2008:244). In both cases, the idiom in question distinguishes between "high-level" and "low-level" domains, in the way we encountered through Connolly's work, noted earlier: namely, where "high" and "low" refer to degrees of complexity without implying an ontological hierarchy. (For Connolly's use of emergentism to allow nonhierarchical distinctions between degrees of agential complexity, see Connolly 2010:23–27.) In both weak and strong emergence, "the high-level phenomenon arises (in some sense) from the low-level domain": the difference between the two is that in weak emergence, "truths concerning that phenomenon are *unexpected* given the principles governing the low-level domain," whereas in strong emergence, "truths concerning that phenomenon are not *deducible* even in principle from truths in the low-level domain" (Chalmers 2008:244).

As this language of phenomena and truths suggests, emergence is largely taken as an issue for epistemology and heuristics; as Dave Elder-Vass explains, however, it is also of considerable significance to reflections on the nature of agency. "The value of the concept," Elder-Vass writes, "lies in its potential to explain how an entity can have a causal impact on the world in its own right . . . that is not just the sum of the impact its parts would have if they were not organised into this kind of whole" (2010:5). Of course, in a human-exceptionalist understanding, this idiom would confine us to the realms of mere efficient causality rather than agency proper; but as we have rejected this understanding, and with it the a priori distinction between levels of agency or causality, we are free to adopt it as a way to account for the appearance within a situation of a new grouping that exhibits capacities whose existence and distribution are not given in advance by the features of that situation. In my terms: from among an array of beings, an alliance has appeared, constituting a subset of these beings as participants and exhibiting agency; that is, functioning as a locus of action. Nothing in the array determined the appearance of this alliance; as we are no longer in the realm of substance metaphysics, the beings in question have no timeless essences whose combination might be known in advance to give rise to such a thing. As a strongly emergent phenomenon, then, the alliance is both irreducible to its parts and causally effective. Its appearance constitutes what Barad calls an "agential cut" (1997:140). Just as in her account of Niels Bohr's quantum mechanics, the act of measurement itself decisively configures the features of a system and installs the distinction between the "object" and the "agencies of observation" (in this case, through the decisive mediation of a particular apparatus of measurement), so the emergence of an agential alliance interrupts its situation with the arrival of a new power and transforms beings into participants, decisively reconfiguring the elements of this situation *in the act* in which it appears.

Crucially, both this appearance and the alliance in question remain wholly immanent to this process of emergence. Yes, strong emergence is also referred to in the literature as *ontological* emergence, but this refers to the arrival in the world of something genuinely novel, not to some irruption from a wholly other scene (see Clayton 2008:7–8). High-level properties are irreducible to and inexplicable by their low-level counterparts—but the distance between them is not the infinite separation of distinct orders of being. For example, as we will see when addressing the theme of the decision, an operative concept of reflexivity will be a necessary part of my overall model. But thanks to the insights of second-order cybernetics, it becomes possible to conceptualize reflexivity without transcendence: namely, as reflection on a process from within that process. We can thus secure the reflexivity needed without having to index this to a faculty of self-consciousness understood as ontologically of a wholly different order, while also understanding this reflexivity as making a drastic difference within its situation. (At this point, we might compare this dynamic with a Hegelian account of the emergence of subjectivity through the self-differentiation of substance—as long as this subjectivity is understood, as it hardly ever is aside from the work of Catherine Malabou, as neither ontologically transcendent in relation to its object nor defined by human-exceptionalist properties. We will return to this in chapter 3.)

The motif that serves here to capture this idea of immanent emergence is the one that gives this book its title: the *accident*. With its associations of random happenstance and an absence of volition, this motif suggests the severance of agency from intention while also emphasizing an absence of transcendence: an accident might well befall us, but it never comes from nowhere. Its elements already given within the situation, it is, perhaps, a less grandiose version of Badiou's Event: not so much the punctuation of history by the void that cannot appear as such, as a twist, an unpredictable deviation that rearranges these elements and thereby makes a decisive change—the Lucretian *clinamen* that composes a world. Given its explicit rejection of intentionality, however, the accident—even as operator of decisive change—would seem to be wholly unsuited to a model of political agency wanting to maintain such apparently decisionistic categories as prescriptive mobilization and commitment. Again, though, as with the concept of agency, this is precisely why I will be using the term. At stake in the following arguments is the question posed by the juxtaposition in my title: can we meaningfully think effective political *agency* as *accidental*? The apparent oxymoron of *accidental agents* signals the need for a venture such as this to reconcile the refusal of sovereign intentionality declared by its first term with the *decisive* effectivity—in both senses of the term—implied by our usual understanding of the second. The motif of the accident thus goes to the heart of the intervention this study proposes: its claim that such a reconciliation is not only possible but essential.

In effecting this reconciliation, the challenge is accordingly to understand accidental agential emergence as the locus of decisive intervention—indeed, of futural commitment. At this point, all the substance-metaphysical baggage of the concept of agency returns once more, as indeed I want it to do: with the decision, commitment, this introduction of a futural dimension, are we not back inside the human-exceptionalist theater of intentionality? This is where we meet the accident as also the site of the decision. Paradoxically, no doubt; but again, the tension between the two terms as habitually understood indicates precisely what is at stake in this venture. If the suspension of the human-exceptionalist machinery of intentionality obliges us to withdraw the decision from its fantasy status as sovereign projection, immersing it instead in process, the accompanying conception of agency as an emergent property maintains an insistence on the drastic change brought about by the emergence of the agent in the act. If nothing could have predicted this arrival—if the agency in question is indeed emergent and, in this sense, accidental—the act in which this composite agential alliance appears will here both make a drastic difference to its situation and comport a futural commitment.

Building out from the respective versions of the concept we find in its two great contemporary thinkers, namely Stiegler (chapter 2) and Malabou (chapter 3), I will use the accident to characterize agential emergence as bringing forth a decisive intervention, in both senses of the word. In this way, I argue that immanent, processual agential emergence also entails orientation toward the future; this is crucial to my case, because it gives decisive future orientation without (cognitive self-) representation, reflexive projection without transcendence. The decision as heir to the accident—but no less decisive for that.

In securing decisive commitment in this way, such immanent futural orientation also allows my model to accommodate the final aspect we need to consider in this introductory sketch: namely, what I have been presenting as the necessarily conflictual side to any thoroughgoing account of politics. Like many, I imagine, I am very happy to agree with Chantal Mouffe and see a desirable politics as a practice of "agonism" (in which adversaries confront each other in a context defined by mutually agreed norms)—the alternative being antagonistic confrontations between enemies who refuse to recognize such common ground and so maintain their hostility in unsublimated form (see Mouffe 2013). In the years since Mouffe first formulated this position, the desirability of such a politics has, of course, only been heightened by the proliferation of powerful actors whose methods are openly antagonistic. Part of the challenge posed by such actors to a broadly liberal tradition has, however, been the struggle they have occasioned *within* the political field over what counts as politics. And if this struggle is indeed situated within the political field, we are, I think, obliged to accept that antagonism does form part of this field. (On these debates, see, for example, Brown, Gordon, and Pensky 2018.) As Mouffe suggests, then, conflict is an irreducible part of politics broadly understood (2013:3–4): although it is certainly preferable for a given political arena to be constituted by a mutual agreement to sublimate antagonism into agonism, such an agreement constitutes the welcome opening of a particular form, not a condition of possibility for any politics at all.

In addition to Mouffe, we can usefully draw here on the work of political theorist Oliver Marchart. In his tellingly entitled *Thinking Antagonism*, Marchart sets out what he calls "the conditions to be met in order for us to reasonably speak about politics" (2018:36). These are six in number: collectivity, organization, strategy, conflictuality, partisanship, and what Marchart terms "becoming-major," which he glosses as the requirement that "politics is geared towards constructing a symbolic majority" (36). If "becoming-major" is directed—antagonistically—at the "becoming-minor" of a Deleuzo-Guattarian micropolitics, we can justifiably see this list as a whole as evoking a resoluteness and a capacity for projection that would typically pull against a conception of agency as other than a distinctively human quality. Despite this, I will take conflictuality and partisanship, at least, as indispensable to any model of political action that aspires to a good degree of descriptive accuracy. Indeed, as I have said, I will do so in part precisely to claim that these dimensions, which I take to be indispensable to an adequate account of political agency, are compatible with an account of such agency that dispenses with all appeals to human exceptionality.

To be clear, I will certainly not be arguing that partisan conflict should account for the whole of what we mean by politics, let alone that agency itself should be understood as always and everywhere conflictual. I have no particular fondness for antagonism and even less for the machismo its invocation often serves to flaunt. With Mouffe, Marchart, and others, however, I do believe that partisan conflict is an irreducible feature of the broad field of political activity. Given this, a model of political agency as distributed that aims to be both complete and effective will have to be able to incorporate a conflictual dimension—which is accordingly part of what I seek to do in this book. Different contexts will motivate differences in emphasis: for the sake not just of descriptive accuracy but also of rhetorical persuasiveness, it will be more or less appropriate now to highlight frontal conflict as one mode of distributed political agency, now to stress the role of the indirect and the capillary. Because I hold that partisan conflict must feature in any adequate account of political agency, and that such conflict has thus far not been much emphasized in work on distributed political agency, the following study will be one context in which the conflictual will be highlighted.

My understanding of political agency will thus accord with that of what Raffaele Marchetti calls the "radical tradition": namely, as participation "in the struggle to define the modalities of life in common," a definition that, as Marchetti points out, "stress[es] the conflictual dimension of politics" (2013:14)—with the key difference that I will be working with a significantly expanded notion of "life in common." And this does make quite a difference! If I hold that our existing accounts of distributed political agency need to find room for the dimensions of conflictuality and partisanship we find theorized in this "radical tradition" (represented here thus far by Badiou, Mouffe, Marchart, and Marchetti), I equally hold that the understanding of agency as distributed set out in these existing accounts can help us to develop a considerably more expansive conception of politics than those typically found in the worlds of political thought and political theory, radical or not. For all the internal diversity of these fields—from the traditions of conservatism, liberalism, communitarianism, libertarianism, and socialism, say, to theories of democracy and civic republicanism, to the *Realpolitik* of Hobbes, Schmitt, or Lenin, or the redistribution of the sensible realm à la Rancière—they remain overwhelmingly committed to the human-exceptionalist understanding of politics we met earlier, in which only humans do politics on the basis of unique, invariably linguistic-cognitive, capacities. As I have stated, I have no intention of arguing in what follows that political activity is somehow indifferently undertaken by humans and nonhumans. On the contrary, if my aim is to look more closely at how political actions are performed by alliances of ontologically diverse participants, this scrutiny brings us not less but more and better differentiation among the respective contributions of these participants. Such more and better differentiation does, however, deliver a model of political action in which politics is no longer undertaken *by humans alone*.

What is more, the defamiliarization of politics enabled by this position of ontological diversity obliges us to revisit what we take to be the essential elements of the political field. As we will see throughout this study, and as I will schematize explicitly in its conclusion, in place of the agora and the town hall, representation and discursive participation, sovereignty, the citizen and the demos, we find alliances and their participants, decisive action and the composite decision, and partisan commitments within a conflict over values. The defamiliarization in question is not total, it is true (no effective defamiliarization ever is); as this reference to conflict over values suggests, the Aristotelian horizon of the good life as the final cause of this activity is still to some extent in place here. And, indeed, some other features of that humanist political scenery will persist in the model I develop in this book—notably, the participatory dimension invoked in the figures of the agora and the demos. But if these features do persist, this will be persistence in the mode of mutation, with drastically expanded stakes; and various others will certainly fall away, most particularly those that serve to frame politics essentially in terms of discursively mediated representation. For the good life being disputed is no longer solely that of the *zoon logon ekhon*, no longer staged within a human-exceptionalist theater of intentionality whose machinery of linguistic-cognitive representation is designed to secure this being's fantasy transcendence. It is now that of beings of all kinds, who are, moreover, involved in its conflictual definition alongside that supposedly exceptional political animal.

The shift in question might thus fairly be thought of as that from politics in a conventional sense to what Latour calls "cosmopolitics," which he frequently defines as the progressive composition of a common world by and for beings of all kinds (see, for example, Latour 2014:35–36), or, simply, "the management, diplomacy, combination, and negotiation of human and nonhuman agencies" (Latour 1999:290). Except that we will have to take seriously the quasi-Leninist formula with which Latour immediately glosses this definition: "Who or what can withstand who or what?" (1999:290) Cosmopolitics, by all means—but with full weight given not just to the measuring Latour's gloss implies, the delineation and evaluation of power differentials, but to the partisan commitments and conflicts in which these differentials effectively appear. Politics, then: as Marchetti's "struggle to define the modalities of life in common" but with antagonistic alliances of ontologically diverse participants as its decisive actors.

OUTLINE

In what follows, then, I develop our existing theories of distributed agency as an emergent property of ontologically diverse alliances, notably by demonstrating the compatibility of distributed agency with decisive intervention and partisan futural commitment. As part of this, as I have suggested, I build an account of the decision as strong immanent discontinuity, reflexivity without transcendence, at once inflection and interruption—of the decision as processually enmeshed, heir to the accident, yet still sharply interventionist. And through this model of decisive distributed political agency, I also offer a general model of politics in which the usual human-exceptionalist, linguistic-cognitive entry requirements are suspended but the ontologically diverse alliances that now form our political actors remain combative figures of partisan commitment and normative prescription. Accidental agents; antagonistic alliances.

To accomplish this goal, the book presents in-depth analysis of its three key thinkers: Latour, Stiegler, and Malabou. If Latour's importance here is evident, Stiegler and Malabou are cited less in arguments around distributed agency. One of the aims of this study is accordingly to move beyond the largely expository or critical modes of their reception so far, to show the broader contribution that their respective philosophies can make. Unlike in most of the existing literature, my approach to Stiegler and Malabou takes from their work not so much this or that concept as, rather, in each case an underlying mechanism, which—duly adapted, as necessary—forms a key element of the model I am seeking to build. Each of my three titular thinkers thus forms the object of a substantial chapter-length discussion in which their ideas are first clearly unpacked before being mobilized by means of close reading of their argumentation. Through this, I establish exactly what I will be taking from their work to develop my model. These chapters are linked by two horizon sections: these are designed as laboratories in which to confront the resources established in the preceding chapter with the demands of particular contemporary geopolitical issues (illegalized migration in the context of climate change for horizon 1; digital-algorithmic politics for horizon 2). The book accordingly develops through a rhythm of critical engagement with its key thinkers (in chapters 1, 2, and 3), interspersed with testing of the results of this engagement and cumulative development of its argument (in horizons 1 and 2 and the conclusion). Its model is built progressively, from horizon 1 to horizon 2 to the conclusion, with chapters 1, 2, and 3 supplying the relevant materials.

As I have said, the book's starting point is the refusal of strong ontological transcendence in the form of human exceptionalism.

In chapter 1, we see this refusal in Latour—but we also see that the consequences of its translation into the realm of politics leave him unable to realize the descriptive accuracy and, indeed, political effectivity he seeks. It is this that raises the book's central motivating question: namely, whether the refusal of strong ontological transcendence that characterizes theories of distributed agency necessarily leads to an ineffective politics, limited exclusively to the post hoc description of processes—or whether, on the contrary, this refusal can be compatible with a decisive politics of prescriptive mobilization and futural commitment.

The first horizon section addresses this question by developing the book's model of agential alliances as emergent from and immanent to their situation and by configuring such alliances as antagonistically engaged. With this section, the first stage of the book's argument is in place: thanks to this possible antagonistic aspect, and its crystallization of the full range of existential stakes composing its situation, the intervention of an agential alliance produces a decisive change in a state of affairs. Horizon 1 also presents a first encounter with normative questions of responsibility: the composite action of the alliance means that responsibility, too, is here composite, both resting with the alliance as a whole and parsed out according to its constitutive local contributions. The participants in an alliance thus emerge as *eminent sites of accountability*, with differentially distributed kinds and degrees of responsibility relative to their position in the action of the alliance. Analysis of this distribution is consequently also the occasion for the first appearance of the *more and better differentiation* of agential participation that runs as a motif through the horizon sections and is picked up in the conclusion.

Thanks to the figure of the antagonistic alliance, then, with its crystallization of the stakes of its situation and its more sharply delineated conflictual interventions, distributed agency is secured as compatible with decisive action—but we have yet to confront the question of the decision itself. This confrontation comes in chapter 2, notably via Stiegler's combination of Simondon's philosophy of individuation with Deleuze's figure of the quasi-cause. By means of this combination, Stiegler shows that drastic change—here in the Simondonian guise of individuation—can also be the locus of decision-making constituted as reflexivity without transcendence and, thanks to Deleuze's quasi-cause, as a processually immanent commitment to a particular value. This constitution is a major contribution to the model developed through the book, in two key ways. First, it is Stiegler who allows us to understand the decision independently of any appeal to ontologically exceptional capacities and who, accordingly, introduces the possibility of distributed agency as decisive in the second sense of the term: namely, as taking a position and engaging a partisan commitment. And second, it is here that the normative dimension of our model really starts to come through—as, indeed, it must—given the emphasis I am placing on partisan commitment as a key part of this model. For with Stiegler, we begin to see just how such a commitment, taking a side within a conflict over value, can be understood as processually immanent.

The second horizon section combines this account of the immanent, partisan decision with the understanding of distributed political agency established in horizon 1, thereby ushering in the book's core model of decisive distributed political agency. Politically decisive interventions are here analyzed as composite acts of distributed decision-making, delineated by means of more and better agential differentiation as made up of contributions at local eminent sites of accountability that remain mutually opaque, and engaging a futural commitment to a particular value within their situation. The second stage of the book's argument is now in place: the agential alliance brings not just decisive change but a decision in the strong sense, which emerges from the conjoint activity of its various participants as a partisan intervention and a committed, normative orientation toward the future.

At this point, the remaining task for our model of distributed political agency is to elaborate in detail how such immanent decisive action can also engage a specifically futural commitment within the time of its emergence. This will be Malabou's signal contribution to the book's argument: her conception of plasticity envisages the possibility of decisive change in a situation devoid of transcendence, and her figure of the "plastic reader" appears in such a moment of decisive change as the emergent location of just such an immanent futural commitment, taking responsibility for the prescription of this value as opposed to that. With the model of decisive distributed agency fully in place, the conclusion summarizes the key dimensions of this model, engages the questions raised by its normative aspects, and brings out further what it offers as an account of specifically political agency.

This, then, is what lies ahead. And so, first we turn to Latour.

1

BRUNO LATOUR

"We Have to Agree to Talk About War"

No one has done more to promote the notion of distributed agency than Bruno Latour. From his work on the anthropology of science to what he called *The Politics of Nature*, through the vast *Inquiry Into Modes of Existence* and, most recently, *Down to Earth*, he has consistently offered the possibility of an alternative, richer conception of agency than any we might find in the human exceptionalism of our dominant philosophies. If these invariably remain wedded to an understanding of effective action as guaranteed only by the supposedly transcendent human will, set against the realm of necessity inhabited by the natural or the machinic, Latour has argued that any account of action as it happens, if it can free itself from substance-metaphysical a priori definitions of what counts as an agent, will find itself describing actions realized by human-nonhuman alliances of all sorts.

Latour's pioneering contribution is doubtless most evident in the success of actor-network theory (developed by Latour principally alongside Michel Callon and John Law, the pair who came up with the term [De Vries 2016:15]), which, in suspending human-exceptionalist claims to ontological transcendence on the basis of exclusive possession of agential attributes such as intentionality, describes agency instead as emergent from complex systems composed of multiple human and non-human elements. Latour's work on the history of pasteurization, for example (Latour 1988), presents a network of alliances between such actors as microbes, scientists, and journals, whose concerted action leads to a new understanding of the range of beings inhabiting the world. From this early work on microbes to his latest writings on the Anthropocene, Latour has pursued this ontology of diverse coexistence, and its related model of distributed agency, with rare tenacity and to great effect. Indeed, if it were not for Latour, it is far from certain that we would be wondering today about the question of distributed agency at all.

Moreover, in terms of the specific project of this study, Latour has consistently given this question a political inflection. His exceptionally wide-ranging, stunningly erudite, and engagingly fluent writings have made it famously difficult to confine him to a single disciplinary pigeonhole. If he has latterly been presented by sympathetic critics as above all a philosopher (see Harman 2009; De Vries 2016), it has been clear since at least the programmatic essay "Irreductions" (in Latour 1988) that, as Graham Harman argues (2014), political philosophy is a key part of this. Not in terms of Latour's own political preferences, though (Harman calls him "a politically benevolent French centrist with progressive tendencies" [2014:18]); rather, in terms of the conception of politics consistent with his ontological commitments and elaborated across several works. Latour has consistently displayed an interest in the nature, modes, and conditions of possibility of politics, understood, in terms close to one of its classical senses as decisions concerning the common good, as the attempt to negotiate the terms of existence in common. (Although in Latour's scenario, the entities engaged in such negotiations are, of course, far more diverse than in any of its classical formulations.)

Although I will be exploring various aspects of Latour's treatment of these issues, however, I am not here seeking to propose an account of something like his philosophy of politics in a global sense. Not that Latour is reluctant to offer definitions that amount to such a philosophy; on the contrary, he has quite a fondness for such definitions, sometimes concluding with categorical statements along the lines of "If this is not possible, then there is no politics" (2005:252). But my aim here is not to extrapolate a general Latourian philosophy of politics, not least considering that Harman has already done so (2014). Rather, I am interested in the relationship in Latour's thinking between his conception of agency as distributed and his understanding of political action: in what we might accordingly call the politics of distributed agency in his work. For Latour's conception of agency as distributed across composite systems already comports a politics, inasmuch as it reshapes our understanding of the nature of the interactions among entities of all kinds, and is invariably presented as carrying normative implications as to the mode in which these interactions might be pursued. We will explore all of this in detail later. Given the considerable influence of his approach to the question of agency, though, it is fair to say at this point that the consistent political inflection Latour's writings have given to this question has played an unparalleled role in establishing the politics of distributed agency not only as a topic worth exploring but, indeed—if we take seriously the expanded, urgent, and contested ecological context that is a major factor in this political inflection—as one of the key political problems of our time.

It is my contention, however, that Latour's essential discussions of this problem take us a good way but no further. Though his devotion to mapping the intricate complexity of tangled alliances has vastly improved our operational understanding of the sites from which agency emerges, the ontological commitment that accompanies this process has, I argue, proved a significant limit to the political contribution Latour's work is able to offer. Latour refuses to entertain dynamics of decisive interruption as anything other than bad fantasies of transcendence: at best, they are a deluded attempt to explain away immanent phenomena by reference to another, truer order and, at worst, the reckless dream of a complete, violent break with the current order.

In this chapter, I aim to show that this desire to keep distributed agency and decisive politics firmly apart leads to key impasses in Latour's work, which significantly reduce its ability to operate meaningfully in the current field as Latour describes it. The chapter will accordingly begin with an account of his model of agency before exploring how this links to political questions in his work. Then I present what I consider the shortcomings in his treatment of such questions—which to my mind leave Latour less able than he might be to honor his own excellent ontological commitments.

MULTIPLICATION AND ARTICULATION

When it comes to the question of agency, Latour recommends methodical doubt. "We do not know who are the agents who make up our world," he writes. "When we speak of men, societies, culture, and objects, there are everywhere crowds of other agents that act, pursue aims unknown to us, and use us to prosper" (1988:35). The watchword here becomes the thought of those scientists who, around the turn of the twentieth century, came across the existence and agency of microbes: "There are more of us than we thought" (35). And in this world of proliferating entities, agency comes to appear as distributed across different agents; or, in Latour's usual term, as hybrid. (I will not be using this term myself, for the reasons outlined in the introduction.) To reinforce the displacement this understanding inflicts on human-exceptionalist claims to agential distinction, Latour's early work sometimes added to the terms *actor* and *agent* that of *actant*, borrowed from the semiology of A. J. Greimas, as a defamiliarizing marker that what matters is not some substance-metaphysical attribute by which some beings would qualify as the subjects of action but *function*. Latour is less concerned with particular terminology, however, than with this displacement itself. As he explains, "I use 'actor,' 'agent,' or 'actant' without making any assumptions about who they may be and what properties they are endowed with . . . they have the key feature of being anonymous figures" (1988:252n11).

With hierarchical assumptions about degrees of agential capacity thus suspended, Latour proceeds to map the distribution of agency. In *Pandora's Hope*, he offers a striking illustration of what this might look like with reference to debates in the United States around gun control. In the polemical terms of this debate, Latour asks whether it is guns or people that kill. Imagine a shooting featuring a citizen holding a gun and the gun's discharge. Who or what carried it out? "Which of them, the gun or the citizen, is the *actor* in this scenario?" Latour answers: "*Someone else* (a citizen-gun, a gun-citizen)" (1999:179). A citizen can hardly kill someone by pointing an empty hand in their face; the gun is hardly lethal while lying on a bedside table. Only together do the two bear this new capacity. Definitively, Latour declares, "A *body corporate* is what we and our artifacts have become" (192). It is crucial to remember, however, that our sense of two preexisting beings (human and gun) having combined to create a third (killer) is misleading here; both of these beings were already, in turn, composed of many other interacting agents—and so on all the way down. As Harman helpfully puts it, Latour "does not say that human and world cannot be separated because they are 'inextricably linked,' but says instead that we should not speak of two such zones at all. Most actors are made of so many human and nonhuman components that it would be ridiculous to assign them to one of these kingdoms or the other" (2009:123–24). And agency, accordingly, is distributed across the composite system in question, whether or not this is a supposedly singular actor.

In this conception of agency, it is, of course, axiomatic that action is not simply the product of a controlling mind or will. "Action is not done under the full control of consciousness," writes Latour, but "should rather be felt as a node, a knot, and a conglomerate of many surprising sets of agencies that have to be slowly disentangled" (2005:44). As a result, "action is *dislocated*. Action is borrowed, distributed, suggested, influenced, dominated, betrayed, translated" (46). Action "does not pertain to any specific site" (60); it is distributed through the field of its production and has to be puzzled out each time. Most recently, Latour has identified such a puzzling-out with the tasks imposed by the Anthropocene as a new geologic era (or what, in the subtitle of *Facing Gaia*, he calls "a new climatic regime" [Latour 2017]).

Of course, Latour is intensely concerned to address the much more familiar existential and political challenges of the climate crisis: I will discuss this aspect of his recent work later in this chapter. As we build up an initial sense of his understanding of agency, however, I will for now focus on this part of these interventions. If the notion of the Anthropocene is usually understood as indicating an era in which human action has become a distinct force of geophysical change, Latour thus interprets it somewhat differently: the challenge of the Anthropocene, as he puts it in *Down to Earth*, is "to *face up to* an enigma *concerning the number and nature of the agents at work*" (2018:77). In his account, the Anthropocene names the moment when what he has long been diagnosing as the fantasy Modern cleavage (see especially Latour 1993 and 2004) between the human realm (of free will, agency, and politics) and the nonhuman realm (of nature, necessity, and an impoverished, reactive, mechanistic level of agency) has been generally recognized as both disastrous and false.

It is as if, in some terrible sense, Latour's time has come. As Christopher Watkin writes, "All of Latour's previous work on the complex relations of humans and nonhumans has prepared him to engage with the Anthropocene" (2017:191). "From this moment on," claims Latour, "everything changes in the way stories are told, so much so that the political order now includes everything that previously belonged to nature" (2017:3). Contrary to the fantasies of an agentially inert external world that have allowed us to set ourselves up as the only true agents, we now discover "agencies that are *no longer without connection* to what we are and what we do. . . . It is as though, behind the phantasmagoria of dialectics, the metamorphic zone were becoming visible once more. As if, under 'nature,' the world were reappearing" (62–63). The "metamorphic zone" in which beings of all sorts share and exchange capacities never went away: we just decided to deny its existence with our fantastical tales of human subjects and nonhuman objects ("the phantasmagoria of dialectics"), redescribing its complexity to what we thought was our advantage.

> When we claim that there is, on one side, a natural world and, on the other, a human world, we are simply proposing to say, after the fact, that an arbitrary portion of the actors will be *stripped of all action* and that another portion, equally arbitrary, will be *endowed with souls* (or consciousness). But these two secondary operations leave perfectly intact the only interesting phenomenon: the

> exchange of forms of action through the transactions between agencies of multiple origins and forms at the core of the metamorphic zones. This may appear paradoxical, but to gain in realism, we have to leave aside the pseudo-realism that purports to be drawing the portrait of humans parading against a background of things. (2017:58)

Concerted action, dispersal, recomposition: this is the rhythm of Latourian agency, and it pulses through the exchanges among beings of all kinds. This pulse is what the Moderns have always denied, creating their delusional but effective fantasy of a gulf of transcendence between Man and Nature—be this human freedom transcending the automatism of nature or nature's determinism transcending human will. With no agent now defined as superior to all others (not even in the mode of care: Man as the shepherd of Nature, *Dasein* as the shepherd of Being), the question arises of how these multiple beings are to negotiate their teeming relationships. The question, that is, of politics.

In an interview from 2012, Latour states that his aim has always been to examine the specificity of politics (Latour 2012b:129). A key part of his engagement with this sphere, he says, has always been the liberation of nonhumans from their "almost total exclusion" by political theory (121). Indeed, according to Blok and Jensen, Latour's engagement with political ecology, for example (which they describe as "clearly articulated in classic liberal-utopian terms") derives from his sense that this offers "a vital opportunity to reformulate a progressive historical struggle for liberation and democracy" (2011:90, 77). The first step in such a reformulation entails an expanded sense of the range of beings in existence: the realization that, in the memorable phrase we saw earlier, "There are more of us than we thought" (1988:35), or what the *Inquiry Into Modes of Existence* calls "an ontological pluralism that will allow us to populate the cosmos in a somewhat richer way, and thus allow us to begin to compare worlds, to weigh them, on a more equitable basis" (2013:21). As Latour puts it in *Pandora's Hope*, then, "No model of political action can be offered as an alternative to the model of the critique until we modify our anthropology of creation" (1999:283). This modification duly made, we can understand that "it is only through an extraordinary shrinking of the meaning of politics that it has been restricted to the values, interests, opinions, and social forces of isolated, naked humans" (1999:290).

And so we take the second step in our progressive reformulation, by addressing the properly political question of how the inhabitants of these plural worlds might engage with one anothers' claims, in the interests of peaceful coexistence. "This is what I take to be the political project of ANT," writes Latour in *Reassembling the Social*, using the common abbreviation for the actor-network theory he developed along with his collaborators: "Once the task of exploring the multiplicity of agencies is completed, another question can be raised: What are the *assemblies* of those *assemblages*?" (2005:260) This is our contemporary *cosmopolitical* task (1993:136): constituting the "new institutions of democracy" (2004:42) that will allow the more or less faithful representation of human-nonhuman composites and within which their various interests can be negotiated. In *We Have Never Been Modern*, Latour presents such an ontologically plural politics as the third great task of democratization, following the inclusion within the modern Western polity first of the citizen, in the eighteenth century, and then, in the nineteenth, of the worker (1993:136). And the horizon of this task is one that might well have resonated with those who were busy expanding the constituency of human parliaments in the eighteenth and nineteenth centuries but which now takes on its full meaning: the recognition and formal constitution of a common world. At one point, indeed, Latour gives as a definition of politics "the intuition that associations are not enough, that they should also be *composed* in order to *design one* common world" (2005:259). What De Vries calls "the lead-question of Latour's cosmopolitics" is, accordingly, "how to establish a common, yet plural world, one in which we are confronted and challenged by—human and nonhuman—others, without taking recourse to some pre-given, agreed-upon authority?" (2016:138). This, then, is the cosmopolitical task for which we find ourselves newly free: "Once we have exited from the great political diorama of 'nature in general,' we are left only with the banality of multiple associations of humans and nonhumans waiting for their unity to be provided by work carried out by the collective, which has to be specified through the use of the resources, concepts, and institutions of all peoples who may be called upon to live in common on an earth that might become, through a long work of collection, the same earth for all" (Latour 2004:46).

As Latour puts it in *Politics of Nature*, (2004:50), this process accordingly entails the two dimensions of *multiplication* (the expansion of the range of beings counted) and *composition* (the question of how they might come together). The task is simply stated: "to imagine a political philosophy for assemblages of humans and nonhumans" (2004:52). Easier said than done, however; because "politics has to get back to work *without* the transcendence of nature" (56), it is now a much more complicated matter of attempting to form the collective without relying on the preestablished coherence created by the founding expulsion of nonhuman beings.

Politics of Nature duly presents Latour's most detailed engagement with political *composition*, in the shape of the bicameral "parliament of things" announced in *We Have Never Been Modern*

(1993:144) and whose procedures this later work describes in meticulous allegorical detail. This model has been criticized on various fronts, notably for its apparently untroubled assumptions regarding the universal desirability of representative democracy as the mode of composition (see especially McGee 2014:120–21; Stengers 2011:366; and Watson 2011:57–66), assumptions that, as Latour's erstwhile collaborators Callon and Law point out (1995), display a strangely modern tendency toward liberal assimilationism. Connected to these criticisms is the concern that Latour's explicit embrace of discourse as the irreducible medium of composition (see, for example, Latour 2004:62) commits him to an "unduly monologic" (Callon and Law 1995:504) anthropocentrism, with human language restored to its substance-metaphysical role as guarantor of ontological transcendence and political agency. And, indeed, were this always a one-way process—humans bringing nonhumans into representation—the risk would clearly be great. But for Latour, all beings are always engaged in translating one another; in broad ontological terms, he describes this as a universal dynamic of *articulation*. If we want to understand how Latour conceives of politics—that is, dealings among all manner of beings seeking acceptable forms of coexistence—this dynamic will be key, for it opens onto his fundamental account of the interactions between these diverse beings and their divergent interests.

The entry for "Articulation" in the online glossary of the *Inquiry Into Modes of Existence* states, "Entities are not dumb, rather they are articulated; we do not speak because we have language but because we conspire with, and participate in, this generalized articulation. It is the articulation of beings that enables us to talk about them. . . . Articulation is not a property of human language but an ontological property of the universe" (Latour 2012a: n.p.). Articulation might accordingly be defined as the relation of association by which an entity enlists other entities through whose translations it comes performatively to exist. In Gerard De Vries's helpful summary, "What an entity *is* depends on its relations with other entities, the web of connections in which it has become established, the translations it has performed and the translations that have been performed on it" (2016:65–66).

Articulation—mediating translation via others—is the key to any existence at all, as Latour emphasizes in the *Inquiry Into Modes of Existence*: "deprived of other beings, any existent whatsoever would cease at once to exist. Its very existence, its substance, is defined by the supreme duty to explore through what other beings it must pass to subsist, to earn its subsistence. This is what I have called its articulation" (2013:454). As Watkin summarizes (using the term *actant* that Latour borrows from Greimas), "Mediation and translation do not designate some unique capacity possessed by human beings alone; they are the way in which every actant exists" (2017:180). Human language is accordingly relativized, repositioned as one local variant of a universal process of mediation and articulation in which entities rely on being translated by others in order to exist. Crucially, though, we must not imagine that some kind of perfect translation, some wholly faithful or transparent representation might be possible if only the translator tried hard enough, paid enough attention. In "Irreductions," Latour already poses as axiomatic that "by definition *faithful* representatives cannot exist, since they say what their constituency has not said and speak in their place" (1988:195). This is no discursive paradise, where goodwill suffices to ensure universally harmonious communication. Rather, it is a universe of *partial* transmission, in both senses of the word: incomplete and skewed. For transmission is inflected by power. And as this suggests, it is here that Latourian articulation takes on a recognizably political aspect.

Latour's critics—including many sympathetic to his project—have often charged him with an undue methodological exclusion of power relations, especially in their sociological form (see, for example, Haraway 1992:332, 2006:139; Barad 1997:58; Harding 2008:35–46; and—from a more sharply divergent perspective—Mouffe 2013:79–82). These criticisms are, I think, well founded; and I will develop my own version of them later. But they should not obscure the fact that Latour's core accounts of how agency is formed are steeped in questions of power. As we might expect, however, power for Latour is not exclusively localized in one kind of privileged actant (whence the methodological rift with established sociological descriptions of certain privileged social actors). Rather, all actants exert power on each other. In the terms of his discursive model, "Every actant decides who will speak and when. There are those it lets speak, those on behalf of whom it speaks, those it addresses. Finally, there are those who are made silent or who are allowed to communicate by gesture or symptom alone" (1988:194).

Within this web of mutual enforcement, some agents emerge as more powerful than others on the basis of their claims to fidelity. "There is only one way in which an actor can prove its power," writes Latour. "It has to make those in whose names it spoke *speak* and show that they all show the *same thing*. Once this is done, then the actor can say that it did not speak itself but faithfully 'channeled' the views of others" (1988:196). Because we know that this fidelity is impossible, however ("Since a spokesman always says *something other* than do those it makes speak"), we have to understand that this claim is an effect of force, which "can always insinuate itself between the speaker and those that it makes speak," and "can always make them say something else" (196). Force in itself is everywhere and as such neutral; the question for Latour is whether or not the more or less forceful associations comprising a given agential network are revealed. When these associations are dissimulated in a performance of supremacy, we are no longer dealing with force but with what Latour calls *potency* (213). If all actants speak for other actants, conscripting them into alliances and silencing them in favor of their own more or less faithful translations, a *potent* force replaces the patient attention that might characterize this process with imposition, dictating the terms of the representation in question and masking its own dependence on those it claims to translate: "A force becomes potent only if it *speaks for* others, if it can make those it silenced *speak* when called upon to demonstrate its strength, and if it can force those who challenged it to *confess* that indeed it was saying what its allies would have said" (1988:197).

It is thus that antagonism arrives in Latour's scenario, as the exacerbation of the low-level conflict that defines his actants' mutual translations. We confront what De Vries calls "a world of a multitude of forces engaged in trials, that may enlist each other to gain strength and that signify their reality by resisting trials" (2016:65). If power is always at work in all directions, some entities seek to dominate this situation: "They wish to act rather than be acted upon. They wish to be stronger than the others" (Latour 1988:167). By themselves, all actants are weak, obliged to make alliances with others in order to act; this follows definitionally from the composite nature of action. The crucial dimension, accordingly, is the *arrogance* of the potent force, its relegation of other actants to supposed passivity as part of its claim to define the nature of the alliance on which it depends (170). So it is that one force dominates others, in what effectively appears as a campaign of colonization: "A force establishes a pathway by making other forces passive. It can then move to places that do not belong to it and treat them as if they were its own" (171).

This gesture of fraudulent domination is so important to Latour that at one point, he even presents its understanding as the core task of his methodology. This aims, he says, "not only at establishing equality . . . but at registering differences . . . and at understanding the practical means that allow some collectives to dominate others" (107–8). As this emphasis on practical domination suggests, the fraudulent quality of potency is no cause for celebration: however illusory, potency is by definition *effective*, taking form only when one force corrals others and claims their efforts as its own. "Since there is nothing but weakness, power is always an impression," writes Latour. "However, this impression is all that is needed to change the shape of things by in*forming* or im*pressing* them" (201). Potency is always, by definition, both illusory and real.

With this account of the nature of interaction in place, and notably the fundamental role it reserves for differential power relations, it becomes possible to ask in Latourian terms the properly political question: what is to be done? More specifically, in the face of the simultaneously illusory and effective nature of potency, what kind of action is required in order to compose the common world? At its simplest, Latour's answer to this question would be: don't act like the Moderns! In slightly more detail, this answer means that in order not to repeat the fatal delusion of believing there to be an ontological divide between humans and nonhumans, any action must refuse to think of itself as intervening *from the outside*; that is, as the transcendence of the currently existing state of affairs. Given that all agents are composites, and all action effected by alliances, no such transcendence is possible.

In *We Have Never Been Modern*, the kind of action to be avoided goes by two names: *critique* and *revolution*. The first of these—which Latour also terms "the modern critical stance," or "the critical project" (1993:11)—designates in its strict sense the attitude of mind that seeks to "purify" nature-culture composites, separating out the two elements and distributing them to either side of a "Great Divide" (12)—not least by its twin fantasies of transcendence, be this human freedom transcending the automation of nature or nature's determinism transcending human will. Latour also extends this sense, however, to take in what he calls "debunking" (43); namely, that intellectual attitude which, mobilizing the movements of both transcendence and dialectical negation, sets itself apart both epistemologically and ontologically from that object whose previously hidden truth it claims to reveal. "Revolution," in turn, might be thought of as the large-scale sociopolitical version of critique: that gesture which sets itself apart from the existing order of things in order both to reveal the corruption of this order and to sweep it away in a time of radical rupture.

In his attempt to wean us off our Modernist attachment to this model of change, Latour deploys two principal arguments. The first is purely contingent, the upshot of recent history: "Today," he writes, "denunciation and revolution have both gone stale" (1993:45). Thanks to what he calls "the miraculous year 1989" (145), the failure of socialism has buried for good the modern fantasy of a definitive break with history, of a new dawn bringing a new order of justice and equality. If this can seem a rather superficial—not to say banal—account of the end of the Cold War (the essay was first published in 1991), Latour's concern is not really with recent history as such. His second argument is that this contingent development has, in fact, allowed us to see only what was always the case: that revolution was never the radical upheaval it took itself to be. "Today," he writes, "that very idea [i.e., "that revolution is possible"] strikes us as exaggerated, since revolution is only one resource among many others in histories that have nothing revolutionary, nothing irreversible, about them" (48). Developing a position first presented in "Irreductions" (1988 165), Latour writes, "Seen as networks, . . . the modern world, like revolutions, permits scarcely anything more than small extensions of practices, slight accelerations in the circulation of knowledge, a tiny extension of societies, minuscule increases in the number of actors, small modifications of old beliefs. When we see them as networks, Western innovations remain recognizable and important, but they no longer suffice as the stuff of saga, a vast saga of radical rupture, fatal destiny, irreversible good or bad fortune" (1993:48).

"If there is one thing we are incapable of carrying out, we now know, it is a revolution, whether it be in science, technology, politics or philosophy" (1993:69). Revolution, for Latour, is *out of time*. Its time has come and gone, and its eschatological self-understanding as a privileged break with previous times can now be understood as a distortion of the nature of temporality, in which "there is no last moment to condemn all those that came before," given that "times are irreducible" (1988:165). There never was any possibility of rupture with our multiple pasts; and if we have finally understood this (after at least two centuries of Modern delusion), we can at last give up on both revolution and counterrevolution, twins conjoined by their attachment to the notion of time's unidirectional arrow, under which any action is conceivable only as a *break*. (With the past, or with the break with the past, respectively, see Latour 1993:67, 69.) Any action must always be a local negotiation within a field of multiple actors and distributed agency, so no break is ever possible, *thankfully*. Despite our persisting attachment to the fantasy temporalities of totalizing progress and definitive change, of a "total upheaval" (1998:4) that would put an end to time itself, this impossibility is neither a "disappointment" (1993:69) nor a "dereliction" (125) but a welcome return to an understanding of the world and of action in it as organized through complex, composite networks. "Let us move on to other things," suggests Latour, before correcting himself. "Or rather, let us retrace our steps. Let us stop moving on" (62). He returns to this thought more fully a few pages later: "What are we to do, if we can move neither forward nor backward? Displace our attention. We have never moved either forward or backward. We have always actively sorted out elements belonging to different times. We can still sort" (76). And this is the nature of the action that is required: namely, the kind that has always been going on (and the more the Moderns tried to "purify" this messy state of affairs, the more densely it grew): "tinkering, reshuffling, crossbreeding and sorting" (125–26).

This refusal of decisive, interruptive action should not be taken to imply that Latour sees no need to take account of conflict, however. We saw earlier that antagonism is a key feature of his account of the relationship between agents: indeed, that the domination of one agent by another is precisely what needs to be understood. As Kyle McGee states, the peace of Latour's common world "is a turbulent, adventurous, and difficult peace: there is no destruction of diversity, tension and antagonism even *within* the common world" (2014:122). In this sense, though it works for a common world, Latourian cosmopolitics is inherently conflictual. As Latour writes in *Facing Gaia*: "descend[ing] from 'nature' toward the multiplicity of the world . . . comes down to reopening the two canonical questions: what existents have been *chosen*, and what forms of existence have been *preferred*?" (2017:36–37) The point, though, is that conflict is to be viewed not as a means to an end, necessary if regrettable, but, rather, as a bone of contention, requiring translation and further negotiation. Using Aristotle's term "entelechies" to refer to agents, Latour accordingly writes, "Entelechies agree about nothing and can agree on everything, for nothing is, in and of itself, either commensurable or incommensurable. Whatever the agreement, there is always something upon which disagreement may feed. Whatever the distance, there is always something upon which an understanding may be built. To put it another way, everything is negotiable" (1988:163).

Again, it is important to stress that this position need imply no facile faith in dialogue: "negotiation," for Latour, can cover the whole range of modes of interaction. "Was it a battle, a ceremony, a discussion, or a game?" he writes. "This is also a matter of dispute, a dispute that continues until all the entelechies are defined and have themselves defined all the others" (1988:164). The baseline insistence that "everything is negotiable" does seek to keep dispute in the mode of discussion, however, and away from the mode of war. In *Pandora's Hope*, Latour writes, "Unfortunately, as we have learned so painfully in this century, wars have devastating effects, since they force every camp to stoop to the level of its adversary. War has never been a situation in which to think subtle thoughts, but rather has always offered licence for taking short-cuts, seizing any expedient at hand, and riding roughshod over all the values of discussion and argumentation" (1999:299).

Promoting discussion over violent conflict (without ignoring the latter), Latour would have us espouse what he sees as the noble and misunderstood patience of the politician, always working out compromises. Indeed, he writes, "It takes something like courage to admit that we will *never do better* than a politician" (1988:210). To seek to burn past this low-level, messy negotiation is to fall back into the Modernist fantasy of radical rupture, and that way lies disaster: "Those who believe that they can do better than a badly translated compromise between poorly connected forces always do *worse*" (211). Mediocre, negotiated compromise has always been the rule, for Latour, and happily so; violent conflict, though real enough, is best understood as an exception to be avoided if at all possible. Rather than soldiers, we should think of ourselves as diplomats, representing the interests of the alliance that we are—bringing these into discussion to work out the least bad compromise available with the alliances that surround us. Thus it is, says Latour in *An Inquiry Into Modes of Existence*, that values are kept alive (2013:481) and it becomes possible to begin "a process of composition that, in order to achieve a common world, is largely political" (2012a: n.p.).

Indeed, in *Inquiry*, he develops this imperative into a full-scale description of the mode of politics, emphasizing its composition of such a common world via laboriously patient negotiation of the fundamental disagreements to which the incompatible concerns of various parties give rise. What matters is to keep the process going, precisely because this is a realm not of harmony but of dissensus, not of continuity but of discontinuity. Good political practice keeps on working at and working around these disagreements, so that ongoing negotiation becomes habitual and each participant "feels obliged to act and able to speak in such a way as to avoid interrupting this paradoxical movement." "When this happens, a *political culture* begins to take shape. . . . *Democracy* becomes a habit" (2013:343). This process can always go wrong, of course, and frequently does; the mistake, essentially, is to fail to relativize a position in anticipation of its qualification by a subsequent intervention. The best way to act politically is not to do this or that: it is to act in such a way that the process keeps going. Good political action is accordingly not only indirect but to an extent self-ironizing: "Being indignant is fine, but preparing yourself to pass on to something else is better. . . . In other words, at each point, the proof that one is not lying is given by what follows in the curve and by the anticipation, the hope, of its necessary return, its renewal, and its future extension" (2013:344).

Extraordinarily, for readers familiar with his previous work, Latour's term in *Inquiry* for the discontinuities between parties that necessitate ongoing negotiation turns out to be none other than *transcendence*. He invokes other figures—notably, hiatus and chaismus—before introducing this term with due acknowledgment of his readers' likely surprise: "Whatever these expressions may be, they all aim to grasp the break, the step, yes, the *transcendence* of the political" (2013:347). Anyone who has read Latour will be more than aware that the obsession with transcendence was the signature failing of the Moderns and that modeling politics without transcendence was the mission assigned to *Politics of Nature*. And yet the notion is dramatically rehabilitated in *Inquiry* as a term for the discontinuity between existents or modes of existence on which Latour has always insisted. It comes here to name just that irreducibility to any synthetic overview that had previously led him to refuse the term. Reintroducing the notion, Latour accordingly splits it in two: just as there is good and bad political action, so there is good and bad transcendence. The entry for "Transcendence" in *Inquiry*'s online glossary helpfully opens by explaining the difference. "Bad" transcendence is what the term previously served to name, now described with added metaphysical baggage as "seek[ing] a foundation in a substance by breaking with courses of action and trajectories"; whereas " 'good' or, better, 'small' transcendence, seeks to focus attention on the hiatus, the discontinuity, the step, the threshold by which all existents must pass in order to subsist" (2012a: n.p.). In general, "small transcendence" is now being used by Latour to name the gap between existents that makes *articulation* via others necessary to their subsistence, "the leaps and the thresholds that have to be crossed one after another in order to exist a little longer" (2013:211). At this point, small transcendence turns out to be everywhere: "there is always a leap, a fault line, a lag, a risk, a difference between one stage and the next, n and n + 1, all along a path of alterations. Continuity is always lacking. . . . Transcendences abound, since between two segments of a course of action there is always a discontinuity of which they constitute, as it were, the price, the path, and the salvation" (211).

Discontinuity, then, crossed and recrossed by attempts at articulation—but crucially, with no change of level. It is this horizontality (which measures only in terms of what separates *this* from *that*, *us* from *them*, *here* from *there*) that differentiates good, small, slight transcendence from its big, reductive cousin, this latter being defined by its dream of elevating to a higher plane in order to provide a more accurate overview of other existents. Axiomatically, "There is aggregation; there is no break in level. There is mini-transcendence; there is no maxi-transcendence. There is piling up; there is no transmutation. There is one level; there are not two" (2013:402). As Latour summarizes, "carefully preserving the distance between the commensurable and the incommensurable, that's good transcendence; extracting oneself from situations to seek the 'external' viewpoint that alone makes it possible to 'judge' situations that otherwise would remain merely 'factual' is the classic example of bad transcendence" (462).

In general terms, then, this is the return of a modest, rehabilitated version of transcendence, whose emphasis on discontinuous articulation seeks to lock out ever more firmly its bad, arrogant avatar. In specifically political terms, the term receives two inflections. First, each political intervention is slightly transcendent in relation to those around it, as discord and discontinuity are here the rule, and the political relay is in a sense the production and maintenance of viable articulations between these moments. Second, political speech must veer off, refuse to remain tied to those it represents, and in this sense become slightly transcendent in relation to them in order to seek the compromise that will allow them to coexist with those others whose interests are also at stake. Latour combines the two as follows, defining "the transcendence of the political" as "what we are sketching in here by the overly geometric notion of *curve* that makes it necessary to *distribute* the little transcendences *all along* what is becoming a Circle, against the temptation to go straight. It will have become clear by now that everywhere there are only little transcendences" (2013:347).

Latour's vision of "felicitous" politics thus entails representatives understanding both the fact of discontinuity and the necessity of articulation (without which there would be precisely no representatives); understanding also their own discontinuity relative to those about to intervene and the need to find circuitous routes by which to arrive at a compromise with these others that might allow the discussion to continue; and being able constantly to relativize their own position so as to allow another to pick up the relay. The political categorical imperative, for Latour, runs as follows: "Speak publicly in such a way that you will be ready to run through the entire circle, coming and going, and to obtain nothing without starting over again, and never to start again without seeking to extend the circle." And he observes, "If there is such a thing as the dignity of politics, the truth of its enunciation, it lies in its having agreed to put itself to the wheel in this way, to have yoked itself to such a grindstone" (2013:349).

At this point, the universal fact of mediation, of the processes of articulation through which any existent at all is translated into being, becomes a value with politically prescriptive force. Latour would, of course, dispute the fact/value distinction as resting on a fantasy external viewpoint that is the very model of bad transcendence. So let us say, instead, that mediation and articulation here shift registers, from a metaphysical one in which the descriptive mode was privileged (however much it clearly already implied a set of normative claims) to a political one in which the prescriptive takes precedence (however much this requires justification through descriptive relevance). As Harman notes, then, "There has never been anything other than mediation, yet at the same time it is good to create *even more* mediation" (2014:44). The shift of register Harman neatly catches here will have considerable significance for Latour's contribution as a thinker of political agency, as we will see. For now, though, this is his portrait of a desirable politics. No total upheaval, no radical rupture, no unmasking of the real interests lying behind miserable maneuvering; just patient negotiation committed to its own perpetuation, so that the work of multiplying composite entities and composing the common world might keep going.

IMPOSSIBLE WAR

Having set out the key aspects of Latour's approach to agency, especially in its political mode, I now turn to a critical analysis of its limitations. Its achievements are, of course, considerable, and what follows is not designed to dispute them. On the contrary, part of my claim will be that Latour's limitations prevent him from securing these valuable achievements as firmly as he might. The welcome displacement of sovereign, strongly transcendent, unique human agency to a model of agency as distributed horizontally through heterogeneous alliances is not Latour's achievement alone, as we saw in the introduction, (Indeed, to be consistent, it could hardly be thought of in these terms.) But his contribution to this displacement has, of course, been immense and immensely influential. What I want to argue now is that his different definitions of good political action, and the ways in which he tends to characterize the political field, end up limiting the effectiveness of his approach and compromising its ability to contribute not only to the greater ontological egalitarianism he seeks but also to the mobilization he calls for in the face of climate emergency.

As we know, then, Latour seeks to guide us away from a "Modern" understanding of politics as the realm of decisive action, of radical ruptures breaking entirely with the existing order of things. In order to do this, however, he risks distorting the political field, polarizing it until it features just its two extremes: the patient compromise and the definitive break. One of his characteristic moves is to reassure us that where we might have thought we faced terrible difficulties, we in fact have nothing to fear. Alluding to the notion of capitalism as an economic system founded on the principle of general equivalence, for example, he writes, "Like God, capitalism does not exist. There are no equivalents: these have to be made, and they are expensive, do not lead far, and do not last for very long. We can, at best, make extended networks. Capitalism is still marginal even today" (1988:173).

Don't panic—what you think of as a daunting enemy doesn't even exist! Latour's tactical overstatement doesn't help him here. Of course, he isn't claiming that capitalism doesn't exist, only that it doesn't exist as the fully realized, smoothly functioning system described by its boosters and, sometimes, its detractors. In its reality, capitalism is, in fact, like anything else, a matter of messy, local compromises and alliances, of weak actors seeking the force of others to make their way as best they can. Latour develops the point in *We Have Never Been Modern*: "Take some small business-owner hesitatingly going after a few market shares, some conqueror trembling with fever, some poor scientist tinkering in his lab, a lowly engineer piecing together a few more or les favourable relationships of force, some stuttering and fearful politician; turn the critics loose on them, and what do you get? Capitalism, imperialism, science, technology, domination—all equally absolute, systematic, totalitarian" (1993:125–26). There is plenty to admire in these arguments. Latour's point is that the enemy will be more effectively opposed on the basis of an accurate assessment of its real force, not the paranoid fantasy of its total power. As he puts it, "In the first scenario, the actors were trembling; in the second, they are not. The actors in the first scenario could be defeated; in the second, they no longer can. In the first scenario, the actors were still quite close to the modest work of fragile and modifiable mediations; now they are purified, and they are all equally formidable" (1993:126).

Latour returns to this point in *Reassembling the Social*, again emphasizing that his aim is precisely to enable possible modifications: "It does not require enormous skill or political acumen to realize that if you have to fight against a force that is invisible, untraceable, ubiquitous, and total, you will be powerless and roundly defeated. It's only if forces are made of smaller ties, whose resistance can be tested one by one, that you might have a chance to modify a given state of affairs" (2005:250). Indeed, in these pages, such low-level intervention becomes the very definition of politics:

> Is it not obvious then that only a skein of weak ties, of constructed, artificial, assignable, accountable, and surprising connections is the only way to begin contemplating any kind of fight? With respect to the Total, there is nothing to do except to genuflect before it, or worse, to dream of occupying the place of complete power. I think it would be much safer to claim that action is possible only in a territory that has been opened up, flattened down,

> and cut down to size in a place where formats, structures, globalization, and totalities circulate inside tiny conduits, and where for each of their applications they need to rely on masses of hidden potentialities. If this is not possible, then there is no politics. (252)

Against potent princes, Latour's recommended strategy is to work on the alliances that alone constitute their force, sapping their power at its roots, and leaving their potency the mythical claim it always was. Again, there is much to admire here. The strategy does indeed sound like an excellent one, not least thanks to its sober evaluation of the real state of affairs. Oddly, however, this position itself polarizes the field. As Benjamin Noys writes of Latour's reassurances that "capitalism doesn't exist," "This treatment of capitalism as a fantasy of total domination at work in the heads of its critics and supporters is achieved by posing a ridiculously high standard for what would constitute capitalism" (2010:87). Either we accept that our adversary is indeed an obviously fantasmatic totalitarian Evil, or we must accept Latour's trembling tinkerers, as if there were no kind of agency between the two. In what world are there no intermediate positions between the "conqueror trembling with fever" and the Evil of all-powerful imperialism? Although Latour elsewhere works to keep the question of effective domination open, not least in the importance within his model of differences in scale and power between actors, when he handles the issue in this way, he tends to foreclose it. He will admit differences of scale only up to a point: if these start to look threateningly disproportionate, they must be either dismissed as nonexistent or dissolved in the black night of totalitarian Terror.

This is where I find myself agreeing with Latour's sociologically minded critics: if his account of the formation of agential alliances in "Irreductions" lays great emphasis on the effectivity of domination secured through the illusion of potency, his prescription for dealing with such drastic power imbalances lets itself imagine that identifying this illusion will be enough to allow the domination in question to be seriously dismantled through vascular adjustments. In Mouffe's terms, this tendency in Latour's thinking forecloses the "hegemonic" character of the composition of the common world (2013:81); namely, the effective dominance of a given order of relations despite the reality of its contingency and inevitable transience.

If this seems close to Latour's approach in much of "Irreductions," in which he models potency as the combination of effectivity and illusion, Mouffe's critique targets this other tendency, already present here but to the fore especially in *We Have Never Been Modern* and *Reassembling the Social*, in which Latour separates the two elements, vaporizing overweening domination as illusion and securing effectivity exclusively for the realm of small-scale construction and disassembly. When he asks, rhetorically, "Is it asking too little simply to ratify in public what is already happening?" his subsequent irony removes all possibilities other than this one: "Should we not strive for more glamorous and more revolutionary programmes of action, rather than underlining what is already dimly discernible in the shared practices of scientists, politicians, consumers, industrialists and citizens when they engage in the numerous sociotechnological controversies we read about daily in our newspapers?" (1993 144). Quite simply, these are not the only two alternatives. To rule out Modern fantasies of total upheaval, Latour offers us a cartoonish dichotomy in which such fantasies appear as self-evidently ridiculous. I am not convinced, however, either that this remains consistent with his insistence on antagonism as an irreducible dimension of interaction or that it provides an especially accurate description of the political field.

In turn, these polarizations work to prelimit the range of political activity to engagement with only those forces that can be significantly compromised by vascular intervention. On the one hand, of course, all forces can by definition be compromised in this way; on the other hand, however, some forces might *in addition* need to be met with other strategies, which are not usefully dismissed as millennial attempts at total rupture. Where would a street protest fit into Latour's schema, for instance? This will almost certainly be part of a campaign including work on what he calls the "constructed, artificial, assignable, accountable, and surprising connections" (2005:252), serving to keep up a regime's illusion of potency; but it is not itself most accurately described as an instance of such work, being directed frontally against the regime on the macroscopic scale of its effectivity. What about a general strike? We could read this as the enforced blockage of the manifold capillaries keeping a state alive, an intervention across multiple "tiny conduits" (252) simultaneously. And it is indeed partly this and is itself sustained by its own capillary network. But such an action also confronts the state as a whole: its efforts work on specific conduits *and thereby* produce a frontal, large-scale attack. As Harman writes, then, Latour's tendency to polarize the political field in order to disallow Modern fantasies of total upheaval can leave his account of this field sorely lacking in descriptive accuracy: "There is a danger inherent in Latour's position that after fending off modern revolutionary claims, he leaves *insufficient* room for sub-revolutionary change that would still be significant or even cataclysmic" (2014:140). Or, in Barad's admirably succinct phrase, "The political field is not limited to the statehouse" (1997:59).

The irony of Latour's account of politics in *We Have Never Been Modern* and *Reassembling the Social* is that in seeking to undo the modern attachment to a fantasy of decisive rupture, he ends up operating a quasi-modernist purification of his own. (On this, see Elam 1999:6.) In order to insist that politics can and should be just patient, local, low-level, messy negotiation (and despite his belief that this model presents a continuum, up to and including violent conflict), he consigns to the flames the vast middle ground of the political field that cannot be confined to the poles of small-scale tweaking and absolute rupture. In a sense, of course, this is simply to level against him a standard criticism of liberalism; namely, that the liberal promotion of discursive negotiation as the highest political value leaves its proponents incapable of addressing all the political action that takes place outside of this mode. As Mouffe writes, "The denial of 'the political' in its antagonistic dimension is . . . what prevents liberal theory from envisaging politics in an adequate way" (2013:3; and see LaVaque-Manty 2002).

If I am criticizing Latour for this undue limitation here, though, it is not just to identify him as the liberal he unmistakably is (albeit one whose ontology expands exorbitantly beyond liberalism's methodological humanist individualism), but for two reasons that go to the heart of my project in this study. First, I do not believe that his model requires such polarization. As I will attempt to show in the following horizon section, I am convinced both that antagonism and frontal conflict can be accommodated within an account of agency as distributed across diverse alliances, without compelling either the return of ontological transcendence or a catastrophic politics of violent overthrow, and that such an accommodation offers a more broadly convincing account of political agency. Second, the difficulties Latour encounters when he has to negotiate the different degrees of force that operate across the political field as a whole significantly compromise his ability to engage the reality of this field in the way he desires. This really comes to the fore in his interventions on the climate crisis, or what in a 2014 essay he calls "Agency at the Time of the Anthropocene." In these works, the consequences of his polarization of this field prove particularly problematic. For here, Latour is moved by the facts of the situation to reintroduce frontal conflict into his thinking—only to find himself forced by his axiomatic rejection of this mode of action to suspend its reality even as he acknowledges its necessity. Precisely because I want Latour's understanding of distributed agency to be compatible with a politics of robust effectivity, I find this a serious shortcoming in his thinking. His descriptions note the reality of confrontation while demonstrating his inability to conceive of this in its actuality. As I say, I will attempt in the following section to supplement Latour's approach with my own model of political agency as both distributed and decisive, even confrontational; but before we get to this, we must take the measure of the problems caused by his own rejection of such a combination.

Latour's accounts of the Anthropocene, and of the kinds of agency it might be thought to require, make some moves that his regular readers might find surprising. For in contrast to his usual emphasis on agential entanglement and advocacy of low-level tinkering, Latour characterizes the Anthropocene in terms of polemics and stark antagonisms. In *Facing Gaia*, high-stakes confrontation is by definition the issue: the face-off signaled in his title pits the human world against the "Earth" (which is to say, the worlds of innumerable other beings). More dramatically still, Latour insistently calls this conflict *war*. By this, he understands two distinct but interrelated conflicts. The first is what has led to the current era being baptized the Anthropocene: namely, the war we humans have belatedly realized we have been fighting against all manner of other beings. Latour calls on Carl Schmitt here to define war as not just any conflict between adversaries but a confrontation between enemies, with no mediating term, in which the opponent is understood as absolutely foreign and as posing an existential threat. (In this definition, Latour is drawing specifically on Schmitt's *The Concept of the Political*. It is worth observing that elsewhere—notably in *The Nomos of the Earth*, *Theory of the Partisan*, and the writings collected as *Global Civil War*—Schmitt offers different parsings of the categories of enmity and war. On this, see Caygill 2013:104–11; and Szendy 2013:23, 34–44.) In Latour's account, the Anthropocene puts a definitive end to our Modern fantasies of Nature as pure, exploitable exteriority, and this in turn removes the third term that had always mediated previous conflicts, leaving us everywhere in a state of war: "It is this disappearance that obliges each of us to take seriously the 'real possibility' of hostilities, even when we are dealing with 'strange' beings whose existence, in the proper sense, we negate, and who can *in their turn*—this is the novelty—negate ours (2017:237–38; translation modified).

The disappearance of our fantasy of Nature might nonetheless also play a beneficial role: if there is no longer a "state of nature" to relativize all our conflicts, then at least the warring parties—here, the heterogeneous alliances that call themselves "human," on the one hand, and all those against whom they like to define themselves, on the other—will finally have to answer questions Latour repeatedly presents as the heart of the politics of the Anthropocene: "thanks to the disputes over the climate and how to govern it, we are asking the political question again in terms of life and death. What am I ready to defend? Whom am I ready to sacrifice?" (2017:227). In Latour's analysis, only such open declarations can lead to the peace that might still be possible. In this new regime, the imperatives are, accordingly, "designate your enemies and delineate the territory you are prepared to defend" (245). The "repoliticizing" of ecology (223)—that is, its reformulation in terms of enmity—can accordingly force human beings to understand the extent to which they are placed in question by the more or less legitimate demands of all those other beings with whom they are once again discovering themselves to be entwined.

But this is only the first panel of Latour's portrait of anthropocenic war. For these humans are also at war with each other, ranged into "climate change deniers," on the one hand, and, on the other, those fighting for serious recognition of the gravity of the global situation. The stark battle lines in question are consequently also drawn between humans: "We understand nothing about ecological questions if we don't agree to be divided over them. . . . Without first recognizing that humans are divided into so many warring parties, no peace will be possible" (2017:245; translation modified). But Latour will go further: these opposing camps are in fact those still wanting to call themselves "Humans," on the one hand (thereby laying claim to the exceptional ontological transcendence that justifies their ongoing exploitation of all other beings; 2017:251 passim), and those striving to align their interests with those of the other inhabitants of the Earth, whom Latour calls either "Earthbound" (in *Facing Gaia*, 2017:248 passim), "Terrestrials" (in *Down to Earth*, 2018:55), or, occasionally, "Earthlings" (2007:74). This intensification of the state of war does not, however, change the imperatives: Latour continues to insist on the importance of open declarations of war aims and of allegiances, and the same questions return. "Now that there is an acknowledged state of war, it is possible for each of the warring parties to be explicit about its *war aims*," he writes. "Tell us, rather, who you are, who are your friends and your enemies, *whom* you are ready to sacrifice to your own happiness, *which* strangers can put you in a situation such that your existence will be negated: (2017:246; translation modified).

The advantage of declaring this state of war is, according to Latour, that it allows and obliges the parties concerned to declare their interests. The challenge to the deniers (the Humans) will be, "Please, delineate what you are defending, what do you think the land is worth, with what other organisms, what sort of soil, what sort of landscape, what sort of industry, what sort of commerce you wish to survive with" (2015: para. 22). And in response, "Well, list all those beings, those agencies you say you can do without. We will do ours. Then we will draw the territories that are under attack, those that are worth defending, those that could be abandoned. Once this is done, we might compare our chances of losing or of winning" (2015: para. 24). If these are the same questions that are at issue for Latour in the first version of anthropocenic war, the difference here is that in this case, we already have some idea of what the answers might be. For "our opponents are more attuned to what is at stake, better versed in what the words 'possession' and 'defense of one's possessions' mean. They, our adversaries, mobilized long ago" (2015: para. 20).

The language of war has moved in and out of Latour's writings over the years. It is very present in his early work—one of his first books was, after all, entitled *Les Microbes: Guerre et paix*—where it sits at one end of what Latour presents as the spectrum of negotiation. Then it fades (having been widely criticized for its macho leanings; see, for example, Haraway 1997:34; and on this criticism, Luckhurst 2006) before returning with a vengeance in *Facing Gaia* as the baseline definition of the strife generated by the vast destruction human beings continue to wreak. (Upon which it is again criticized on the same grounds; see Haraway 2016:40–43.) And yet Latourian war is never quite what we might expect. If Latour repeatedly describes the Humans as engaged in a "land grab," for instance (e.g., 2015: para. 20), his insistence on their aggressive, acquisitive tactics sits oddly with his account of the clash of which these are supposedly a part. For, as we have seen, the conflict Latour envisages is essentially *discursive*. The face-to-face takes place not, in fact between, Schmittian enemies but between adversaries, sharing a common language. As this suggests, it is essentially a form of negotiation, producing a verbal challenge ("Please, delineate what you are defending," etc.), then consideration of comparative positions ("our chances of losing or of winning"), and then . . . the possibility of peace! Granted, this last is described as coming only after "war has exhausted the warring parties," negotiated by camps that have "exhausted all other options" (2015: para. 25); but it would be fair to say that Latour does not exactly exhaust all options in his depictions of this clash. For all his stress on the *realpolitik* of land grabs and so on, these give precious little sense that anything other than verbal blows might be traded in this supposed "war for the definition and control of the Earth" (2015: para. 21).

Repeatedly, Latour summons the image of conflict before speeding ahead to the discursive negotiations between the incommensurable parties, or proposing the terms on which—despite supposedly being at war—they might prepare for or pursue their ongoing negotiations. Blink, and you might miss it: as bloody strife is invoked but not dwelt on, all the emphasis gathers on the preliminary declarations and eventual peace talks. Repeatedly, war is reduced to the most ephemeral of existences, simultaneously invoked and suspended, or sandwiched fleetingly and reluctantly as the *discursive* transition between two regimes of peace: "What we lose on one side . . . we shall perhaps regain on the other, provided that we *agree to pass from a regime of apparent peace to a regime of possible peace*. Between the two, it's true, there's no point pretending otherwise, we have to agree to talk about war" (2017:226–27).

The language of war hypes up this sense that there is no alternative; but all this turns out to mean is that we will indeed have to use this language.

Reading Latour on climate change, we navigate this circle time after time, sent around to the pole of violent conflict, only to be redirected back up to the pole of discourse. "What to do?" asks Latour in *Down to Earth*, channeling Lenin. "First of all," he replies, replacing the specter of decisive Leninist intervention with Latourian cosmopolitics, "*generate alternative descriptions*. How could we act politically without having inventoried, surveyed, measured, centimeter by centimeter, being by being, person by person, the stuff that makes up the earth for us?" . . . What must be documented are the properties of a terrestrial . . . by which it is possessed and on which it depends, to the extent that if it were deprived of them, it would disappear. The challenge obviously lies in drawing up such a list" (2018:94–95). Back here on the warming Earth, one could be forgiven for concluding that if our political action is to be Latourian, it will indeed be a cold day in hell before we see any such thing—let alone anything of the mobilization and conflict he elsewhere claims make up our infernal reality. The Schmittian question of life and death posed in the confrontation with an absolute enemy has morphed into a familiarly Latourian research program. Tellingly, and with uncanny synchronicity, the inspiration for Latour's proposal here—namely, the "*cahier de doléances*," or "ledger of complaints" compiled throughout France from January to May 1789 (2018:96–97)—was revived in 2019 by the Macron government as part of the "*grand débat national*" it sought to initiate in response to the uprising of the *gilets jaunes*. (Subject to a shift from the *cahier de doléances* model to the more Latourian "*cartographie nationale des controverses*," the initiative was duly welcomed by Latour in an article quite justifiably parsing the uprising as confirming the theses of *Down to Earth*; see Latour 2019.) Far from a way to understand militant action in a conflict produced by the general disorientation of climate change, then, this is at best a peace proposal formulated by an observer and at worst a way for the dominant powers to defuse conflict in the hope of restoring order.

As used by Latour, then, "war" turns out to mean not actual conflict but the declarations that demarcate the parties in question before or after the fact. Thus, in *Down to Earth*, although the war in question is supposedly already under way (it has already been both declared [2018:84] and triggered [3]), it somehow also becomes—with a kind of nightmarish, absurd recursivity—a war about where this war itself *will be* fought. "By pulling out of the Paris Accord, Trump explicitly triggered, if not a world war, at least a war over what constitutes the theater of operations. 'We Americans don't belong to the same earth as you. Yours may be threatened; ours won't be!' (3). Despite allegedly having been "triggered," this supposed war next emerges as only speculatively a matter of armed conflict, as Latour suspends its military dimension in the realm of the hypothetical: "The political consequences, and presumably the military consequences—or in any case the existential consequences—of what the first President Bush had predicted in 1992, in Rio, have thus been spelled out: 'Our way of life is not negotiable!' " (3).

This is politics, not war. As a declaration of what will and will not count as matters for discussion, it sits nicely in the middle of Latour's spectrum of negotiation. For Latour, in fact, being in a state of war "doesn't necessarily mean that we are going to fight" (2017:237, translation modified)—which is, to put it mildly, a curious definition! As if aware of the problems created by his inconsistent use of the term, late in *Down to Earth*, Latour tries to pin the blame on the nature of the war in question: "We are at last clearly in a situation of war," he writes, "but it is a phony war, at once declared and latent" (2018:90). Unfortunately, this move succeeds only in emphasizing that the only war Latour can envisage is war of the phony kind. Time and again, Latour either slips backward (replacing being *in* a state of war with choosing a side in the war *to come*), suspends as merely imminent the war he also suggests is currently under way (by characterizing the situation instead as one of ongoing negotiation), or rushes forward (toward the peace talks). War in Latour is so rapidly elided that one wonders why he uses the term at all, given how uncomfortably it sits with his usual emphasis on politics as piecemeal, low-level tinkering, the patient and fragile construction of alliances. The answer would seem to be that it allows him to mark the gravity of the situation, defined as it is by starkly violent forms of antagonism. His almost instant elision of the reality of such conflict, however, provokes two conclusions. First, although Latour knows that Schmittian existential conflict is the state of affairs, he wants to encourage us toward negotiation. Though this is, of course, a worthy goal, it leaves Latour invoking a level of conflict he cannot bring himself to envisage. Whence the second conclusion: although he knows he needs to, Latour just cannot think the reality of such stark conflict.

Now, on the one hand, of course, these slippages can be taken as so many rhetorical attempts on Latour's part to shift the dominant mood of the times away from conflict and toward discussion. On the other hand, however, to understand them as no more than this would be to miss what they disclose about the kind of politics that is compatible with a strictly Latourian approach. There are, in fact, two fundamental reasons for Latour's inability to stay with the reality of conflict, and both constitute problematic political translations of his core ontological commitments.

The first of these reasons is simply stated. If the watchwords are "more mediation!" or "keep the discussion going!," then, by definition, these watchwords will be inoperative in times of war. If politics for Latour is all about the diplomatic skills of negotiation, the state of war he invokes to describe climate conflict can be only an interruption of politics. As Latour writes in the *Inquiry*, "bringing together principals who are worried enough to trust themselves to the tribulations of diplomacy obviously presupposes a situation that is no longer that of war" (2013:484). Mediation will be reestablished with the end of war and the return of politics, but for as long as we are in a state of war, mediation and articulation are, by definition, suspended as effective governing values. If Latour cannot envision the antagonistic conflict he knows is real, the first reason is that in such conflict, the key terms of his political ontology have no purchase.

The second reason for this impasse in Latour's attempt to think climate conflict is found in his related refusal of drastically interruptive action. As we have seen, although he generally thinks by means of careful differentiations, Latour has a tendency to polarize the field of political action: low-level adjustments at one end and revolutionary total upheaval at the other. And this polarization is designed to disqualify the latter, as relying on an eschatological principle of politico-temporal transcendence that makes it of a piece with the human-exceptionalist claim to ontological transcendence. As we have also seen, however, this leaves his descriptions of the field of political action somewhat lacking in accuracy. Earlier I focused on their exclusion of forms of action—a street protest, a general strike—that do not fit well into either pole. When it comes to an avowed state of war, however, a different problem emerges: that Latour's political thought literally has no place for violent action other than the pole of obviously fantasmatic totalitarian Evil. And this place is constructed precisely in order to disappear, vaporized along with its claims to strong transcendence, leaving all the field to vascular modification.

No wonder, then, that Latour repeatedly slips from the reality of war to its preparations or aftermath; in his schema, its arrogant, potent violence simply cannot appear. This is the consequence of Latour's omission from politics of all those forms of frontal conflict that exceed vascular modification while remaining this side of Terror. If he cannot think the confrontation he repeatedly names as "war," it is because he has excluded from consideration the great range of strongly conflictual action that lies at the violent end of the spectrum of interaction. The irony is that his allergy to anything resembling the claims of "bad" transcendence leaves Latour unable to conceptualize the conflict to which he, better than anyone, has shown those disastrous claims to lead.

* * *

Discursive identification is not going to short-circuit violence into discussion; and sacrifice is a question not of notional alignment but of currently enacted policy. We have become only too familiar with fossil fuel sacrifice zones, from the Niger Delta to the Marshall Islands to the Alberta tar sands, where an unchanging environmental racism sees forms of human and nonhuman life discarded as negligable externalities and a new frontierist extractivism scratches profits from the tapering seams of what Jason W. Moore calls the myth of "cheap nature" (Moore 2015). As Sudanese diplomat Lumumba Di-Aping most dramatically pointed out (at the United Nations Climate Conference in Copenhagen [COP 15] in December 2009), sacrificial climate politics is already a fact. In being asked during those negotiations to agree to a global average increase in temperature of 2 degrees, many African nations were being asked to accept a rise of 3.5 degrees and so to embrace their own sacrifice for the sake of economic business as usual. In Di-Aping's words, "We have been asked to sign a suicide pact," which would legitimize "climate genocide" (cited in Lahoud 2014:512–13; and see Klein 2014:276). Or, in the words of Mary Ann Lucille Sering, climate change secretary for the Philippines, "I am beginning to feel like we are negotiating on who is to live and who is to die" (in Klein 2014:276). Negotiations, yes: not after conflict, though, let alone in its stead but, rather, as already, actively and fatally, antagonistic. In both the specifically military sense and the broader sense of a conflict over ecology whose stakes are existential negation (and mostly in both senses at once): violent confrontation is already here, in the realities of what Latour rightly calls the "full-scale war against all the others" waged by the climate change deniers (specifically, here, their commander-in-chief; 2018:23).

Currently, ongoing military activity serves the geopolitical purpose of securing the short-term future of finance capitalism—along with its resource-intensive, fossil-fuel-driven consumerist front—in the context of drastic climate instability (not least by profiteering from this instability, as Naomi Klein [2007] and Antony Loewenstein [2015] have shown). And, of course, Latour knows this. In *Down to Earth*, for instance, he writes of "millions of people . . . who are driven by the cumulative action of wars, failed attempts at economic development, and climate change, to search for territory they and their children can inhabit" (2018:4). And the war is indeed against all the others: the nexus of warfare, economy, and climate enforcing such human migration is inseparable from the planetary ecological sacrifice known as the sixth mass extinction event. To reiterate, my point is not that Latour is unaware of the reality of these contemporary conflicts. Rather, it is that although he wants to, he is unable to think this reality, and that this inability results from what he takes to be the consequences of the ontological commitments underpinning his account of agency as distributed.

My aim here, then, is to supplement Latour's account by finding a way to model agency as distributed that does not require us to bracket the reality of antagonistic conflict. I have no interest in celebrating such conflict; like Latour, I much prefer negotiation to confrontation. But I also agree with Latour that if we want our understanding of agency to retain much by way of pertinence, this understanding will have to include modes of conflict presented by situations such as the one in which we, along with the other inhabitants of the planet, currently find ourselves. If in one sense I agree with various of Latour's critics that his political writings too often fall short when it comes to envisaging effective action, I certainly do not agree with the common charge (seen in the introduction) that this is programmed by his refusal of human ontological transcendence on the basis of exceptional agential capacities. As a result, my aim is to reconcile our existing descriptions of distributed agency with the need in any functioning conception of politics for the dimensions of prescription, commitment, and antagonistic confrontation. To make good what I have presented as the lacuna in Latour's position by building an account of political agency as both distributed across alliances between actors of all kinds and, if necessary, frontally conflictual. To take seriously Latour's language of existential threat as accurately identifying conflicts already under way, then, and to propose an account of militant distributed agency that might better "stay with the trouble," in Donna Haraway's words (2016). The following section begins that work.

Horizon 1

ANTAGONISTIC ALLIANCES

The question we will address in this section could be described as the possibility or otherwise of militant, partisan mobilization on the basis of an egalitarian commitment extended beyond the form of life we call human. So I propose to start with a slogan. "We are not defending nature: we are nature defending itself!" From Paris in 2015 and Brussels in 2016 to the French ZADists and Greta Thunberg's school strikes, the human activists making this claim are rejecting the hierarchy implied even by the role of guardian and identifying themselves as acting not on behalf of but from within a broad ecological continuum. (On this slogan, see Bineau and Chopot 2017.) In so doing, they are clearly making a claim about the nature and location of political agency, and this claim gives rise to an obvious question. Can human climate militants really claim to be acting as or within a continuum of nonhuman and human beings?

To exacerbate the stakes, we could reverse the terms of this question: can they meaningfully claim *not* to be doing this? Now we have a polarization and a political decision to make—*the* political decision, in fact, which forms the heart of this book.

Do we want to reserve meaningful agency for human beings precisely in order to underpin action in the face of ecological emergency? Or do we want to see this belief in human exceptionality as itself a factor in producing this emergency and so expand our sense of meaningful agency to embrace the conjoint action of all manner of beings? As we saw in the introduction, the first of these positions has no difficulty articulating its translation into politics, political action itself having invariably been defined as a subset of the agency proper only to human beings on the basis of various capacities thought to belong uniquely to these beings (or to some of them, at least). Rejecting this exceptionalism, the second has duly been criticized for its supposed lack of political effectivity; and if I generally consider these criticisms to be ill-founded, as I argued in the introduction, I agree that we still need an overall account of political agency in terms of distributed agency. This account will require us to describe effective political action itself as composed by the conjoint contributions of heterogeneous participants, and to go beyond Latour's tendency to elide actual conflict, as seen in chapter 1, by understanding such distributed political agency as also manifest in the mode of confrontation.

This section seeks to formulate the initial elements of such an account. In doing so, I turn first to the work of the research agency Forensic Architecture, through whose case studies and essays I develop the model of agency sketched in the introduction; namely, as distributed across what I will call *antagonistic alliances* formed within arrays of heterogeneous beings. Developing this model will also entail expanding on a key claim I made in the introduction: that understanding agency as distributed implies not the abandonment of meaningful distinctions between actors but, on the contrary, more and better differentiation. This will also allow us to pursue some initial reflections on the conception of responsibility at work in such an understanding, ahead of the fuller development of this theme in the chapters to follow. After further consideration of the structure of the antagonism that sees alliances confront one another, in which I draw especially on the work of Timothy Mitchell, I conclude by proposing an understanding of political agency as not only ecologically plural but also robust enough to fight for this plurality in the struggles that increasingly define the contemporary planet.

As I say, this will provide, at best, the initial elements of the account I am looking to develop. In particular, subsequent chapters will be tasked with building a model of the decision (and so, in political terms, of commitment and prescription) as processually immanent. But if we will have to wait for the decision, by contrast, this section will formulate a model of distributed agency as *decisive*, in the sense of making a major difference to a situation. By mobilizing this first sense of decisive action, I want to argue against the human-exceptionalist claim that all we have in such cases—that is, absent self-conscious human intentionality—is efficient causality (see Hornborg 2017; and Malm 2018, as discussed in the introduction). I hope to show in this section that the way in which antagonistic alliances bring into play the full range of existential stakes characterizing their situation means that their activity can and should meaningfully be situated beyond mere causality. For all that they will not yet be thought as sites of decision in the strong sense, they will emerge as sites of decisive intervention, including in the mode of antagonistic confrontation. By the end of this section, then, we should have established a model of political agency as both distributed and decisively confrontational.

COMPOSITION, CONSCRIPTION, DIFFERENTIATION

Forensic Architecture was established as a consulting research agency in 2011 at the Centre for Research Architecture at Goldsmiths, University of London. The approach developed by the agency is especially pertinent to our concerns in this study inasmuch as it draws consistently on models of distributed agency and antagonistic alliances, both in describing the situations addressed and theorizing the researchers' own practice. If much in the following account of their work will be familiar from existing models of distributed agency, then (especially in its emphasis on alliances between and conscription of different participants), what Forensic Architecture adds is an explicit understanding both of the situations they engage and of their own activities as such agency in specifically *combative* mode.

Accordingly, it is especially significant here that Forensic Architecture has often analyzed cases in which existing legal frameworks struggle to establish clear lines of responsibility. I will focus here on two areas of the agency's work set squarely within the context of Anthropocenic conflict, in which this problem comes clearly to the fore: namely, its respective engagements with the legal framing of environmental degradation and the policing of illegalized migrants in the Mediterranean. In each case, two alliances of human and nonhuman actors confront each other; and what we discover through these confrontations, I hope to show, is that distributed agency proves wholly compatible with decisive, antagonistic intervention.

In both of these areas, Forensic Architecture's analysts draw attention to the difficulties involved in attempting to determine legal responsibility. For example, working with residents of the Alaskan island of Kivalina in their attempt to sue "twenty-three of the largest oil and gas companies in the world" for the ongoing erosion of their island's coastline as the result of climate change, they note, "In the case, the petitioners struggled to establish direct causal chains of responsibility due to the fact that, by nature, climate change is a distributed and complex process, spanning the whole width, depth, and molecular structure of the earth, sea, and sky" (Modelling Kivalina 2014: 701). (The case in question was brought before the Ninth Circuit Court of the United States in 2012, as *Native Village of Kivalina v. ExxonMobil Corp.*, 696 F.3d849, 856–8.) Rather than simply observing this as a difficulty, Forensic Architecture argues that in such cases, the legal framework in question is not neutral but actively serves to impede the successful prosecution of racialized environmental injustice. That is, the insistence that where there exists no clearly recognizable legal person (as either perpetrator or victim), there can by definition be no offense is itself a construction of the situation whose effect is to withdraw from legal consideration processes defined by long causal chains and complex networks of actors. Discussing moves in Ecuador (2008) and Bolivia (2010) to give legal recognition to "Pachamama" (the pan-Andean deity usually translated as "Madre Tierra," or "Mother Earth"), for example, Paulo Tavares quotes environmental lawyer Christopher Stone's observation: "We are inclined to suppose the rightlessness of rightless 'things' to be a decree of Nature, not a legal convention acting in support of some status quo" (Stone 2010: 2; quoted in Tavares 2014: 565).

Across much of the work of Forensic Architecture, the key "legal convention" to emerge in this sense is the location of responsibility within the sovereign, autonomous, self-conscious human individual agent—or the group or corporate entity conceptualized as analogous to such an agent. In such cases, we encounter the legal consequences of accepting or refusing the notion of distributed agency—and as we will see, these encounters suggest that far from upholding the cause of the powerless by maintaining a functioning model of responsibility, the assertion of sovereign agential personhood more often serves to remove the multifactorial processes constitutive of Anthropocenic violence from legal purview and so to facilitate their perpetuation. By contrast, mapping the responsibility at work in such processes across arrays of conjoint actors, far from blurring the picture, brings this responsibility more sharply into focus.

In the following section, I will first explore how the researchers at Forensic Architecture use the notion of distributed agency to understand the attribution—or, more often, the obfuscation—of responsibility within the processes they analyze. I will then show how they understand their own practice as itself an instance of such agency and, considering it in terms of the Latourian problematic of "speaking-for," will offer my own account of this practice as exhibiting distributed agency in directly confrontational mode, along with the sharper delineation of responsibility that this can allow.

We will begin with illegalized migration against the backdrop of climate change. In March 2011, seventy-two sub-Saharan migrants left Tripoli by boat for the island of Lampedusa. Their vessel ran out of fuel and drifted for fourteen days, during which time the migrants had no food or water. All but nine of the migrants ultimately died. Over the course of their ordeal, they had contact with various other occupants of the maritime space in question (which, in addition to fishing vessels, was heavily populated by military ships and aircraft enforcing an embargo on Libya during its civil conflict). Despite the variously recognized obligation to come to the aid of those in distress at sea, none of these others intervened; the case became known as that of the "left-to-die" boat.

Forensic Architecture produced a report on this case in support of the coalition of nongovernment organizations that were formed to establish accountability for these sixty-three deaths. In the work that provided the basis of this report, the researchers in question significantly insist that such accountability can be established only by an approach that understands agency as both distributed and conflictual. Maritime governance in the Mediterranean, they argue, is marked by "a form of violence that is diffused and dispersed among many actors and which often, as in the case we have investigated, operates less through the direct action of a singular actor than through the interaction of many" (Heller and Pezzani 2014: 659). "The image of the Mediterranean that emerges," the authors conclude, "is that of an environment crisscrossed by 'a thick fabric of complex relations, associations, and chains of actions between people, environments, and artifices.' It is the totality of this field of forces that constitutes the particular form of governance that operates at sea" (678; citing Weizman 2012: 6).

It is the "totality of [the] field of forces" that acts; as in Latour's core scenario of distributed agency, agential alliances emerge as some of the beings making up the array defining a given situation become participants by joining with or, indeed, conscripting, others. The geopower of the sea is not an inherent capacity but an emergent property of the alliance into which it finds itself conscripted: here, an alliance that "turned the sea into a deadly liquid" (Heller and Pezzani 2014: 673). In this case, governments, the European Union, people smugglers, desertification, maritime law, militia profiteering, surveillance technology, the military, civil conflict, poverty, and the Mediterranean itself form an alliance whose composite action issues in death—the Mediterranean as "mass grave" (more than 3,300 deaths in 2017, for example, the majority of these close to the North African coast), "the theatre of a new kind of war" (Kavelakis 2018: 19–20; and see McIntyre and Rice-Oxley 2018).

The descriptions provided by the Forensic Architecture researchers thus characterize the material and conceptual horizons of their activities in terms of the distributed agency that is operative across opposed groupings of conjoint participants. But this is not limited to the situations giving rise to their interventions; for they also understand these activities in such terms. Their challenge to these situations thus itself constitutes the mobilization of various participants within a combative alliance. At the basis of their practice is an understanding of forensics as itself already the activity of conjoint human and nonhuman actors. As Eyal Weizman explains, "Forensic speech is traditionally undertaken as a relation between three elements: an object or building 'made to speak', an expert who functions as the translator from the language of objects to that of people, and the forum or assembly in which such claims can be made" (Weizman 2017: 67).

The work of Forensic Architecture thus entails soliciting the testimony of nonhuman witnesses—although we should note that this is not solely (indeed, perhaps decreasingly) the work of human experts, but it is also undertaken by those technological agents (cameras, software, ultraviolet light, and so on) through which these witnesses deliver their testimony. In the case of the Deepwater Horizon oil spill of 2011, for example, the images constituting evidence of the event were produced by "a remotely controlled underwater camera whose primordial objective was not to mobilize public empathy but to perform strictly technical functions, such as aiding in the assessment of the amount of oil that was leaking and shutting down the hole" (Tavares 2014: 560). Before questions of responsibility and consequences are even broached, the event in question can be brought into representation only as the encounter between machine vision and the protocols of articulation that are operative across what is now its significantly expanded field of relevance. With the event duly established, alliances of advocates, technoscientific interpretative procedures and object witnesses can make their arguments about responsibility and consequences. In the case of Deepwater Horizon, "science is . . . called into court to interrogate the opaque testimony that was gradually recorded in soil transformations, the earth samples functioning as a model of an entire socio-ecological dynamic constructed over decades. The mud registers the historical agency of multiple forces and actors—the impact of the technology used to extract oil, the negligence of state institutions and corporations, the misfortune of migrant peasants, and indigenous communities, wildlife refugees, polluted water streams, and contaminated atmospheres—out of which a complex political history can be narrated" (Tavares 2014: 561). Mud appears in court as a nonhuman witness, "interrogated" not by lawyers alone but by lawyers working with (or against) frameworks and instruments of scientific interpretation.

Returning to the case of the left-to-die boat, the researchers there mobilized "an unorthodox assemblage of human and nonhuman testimony," in which the sea emerged as "a vast and extended sensorium, a sort of digital archive that [could] be interrogated and cross-examined as a witness" (Heller and Pezzani 2014: 673–74). In this model, contestatory forensics pits its own alliances against those whose interests are served by the withdrawal of these scenarios of dispersed action from the purview of meaningful legal consideration.

If nonhuman beings thus "inhabit the courtrooms of national and transnational forums as potential witnesses" (Tavares 2014: 562), we should nevertheless guard against the temptation to see such participation as a free-for-all, in which the capacities of particular actors are dissolved in an amalgam of infinite equivalence. To claim that no actor is effective alone, outside of this or that alliance, and that no actor is endowed a priori with given capacities that it exercises independently of the affordances of its relations, is not to claim that capacities are indifferently distributed across alliances. By itself, mud makes no claim to meaning; without the action of the mud as historical record, however, the claims of its interpreters are empty. The sea alone is not responsible for the death of sixty-three illegalized migrants; without the sea, however, the governmental security array must find alternative participants if it is to be effective.

As Weizman writes, then, when discussing the "material sensibility" through which buildings record the events and processes in which they have participated, "This is not to make an anthropomorphic point: buildings sense not in a human, but rather in a building sort of way" (Weizman 2017: 54). It is a matter not of frittering away the capacity for meaningful action by parceling it out among so many kinds of actor that each receives a vanishingly small share; rather, it is a matter of understanding meaningful action as resulting only from the alignment of diverse participants in alliances across whose multiple locations their capacities emerge as differentially distributed. And, crucially, the capacities attached to particular locations within the alliance do not exist prior to the coalition in question. (Other capacities—such as the capacity to enlist others to a cause—certainly do, but they are themselves effects of existing positions in other alliances.) As I argued in the introduction (drawing on Karen Barad and Katherine Hayles), the combination of emergence and differentiation that thus characterizes an understanding of agency as distributed requires us to give full weight to the implications of that adjective. If capacities emerge as distributed across agential alliances, they are distributed differentially, just as Amartya Sen (1997) describes the differential distribution of capacities as a social good. One consequence of this is that, far from obliging us to throw the baby of responsibility out with the bathwater of human exceptionalism, understanding capacities as differentially distributed across more or less combatively engaged alliances in fact allows us to intensify our sense of the responsibility operative in any given scenario by mapping more sharply the contours of this distribution and the eminent sites of responsibility that they trace. Not less, but *more and better* differentiation: as Jane Bennett puts it, a model of distributed agency "does not thereby abandon the project of identifying (what Arendt called) the sources of harmful effects. To the contrary, such a notion broadens the range of places to look for sources" (2010: 36–37). It is accordingly highly significant for our purposes that the researchers at Forensic Architecture understand their activities in exactly these terms: this is the first occasion we have had to see such more and better differentiation in action.

One good reason for moving away from the grounding of agency in the sovereign individual person (including the corporate legal person) emerges from the examples of diffuse violence whose occultation Forensic Architecture repeatedly challenges: this construction of agency and responsibility works to withdraw some processes from appropriate scrutiny. As Heller and Pezzani write with reference to the case of the left-to-die boat, "As a consequence of this form of systemic violence, the specific responsibility for deaths and violations at sea is difficult to detect and prove"; more bluntly, "the international legal norms established to determine responsibility for assisting those in distress at sea have been used precisely for the purpose of evading and deferring this responsibility" (2014: 659, 670). Faced with a situation in which agency is quite plainly distributed across both space and time, our conventional attachment to punctual individual agential sovereignty plays the role of useful idiot, at best unwittingly complicit in the continued perpetration and occultation of injustice. Against this, then, an analytical method already committed to grasping agency as distributed works in the direction of more, not less, responsibility; it is only thanks to such a method that responsibility can be determined here at all.

Working in this way thus offers what one of Heller and Pezzani's section headings calls "Renewed Opportunities for Accountability" (672), reinscribing responsibility in a scenario from which it had been strategically evacuated. The authors accordingly describe their methodology in terms of a dialectical inversion, or perhaps even a judo move, using the opponent's force against them: "While the fragmentation of juridical regimes at sea so often allows for the evasion of responsibility, in this case it was mobilized strategically towards the *multiplication* of potentially liable actors and of forums where they could be judged and debated" (678). That is, the distribution of agency has no necessary allegiance to the suspension or dilution of responsibility but can, on the contrary, present its intensification, multiplication, and internal differentiation. As Adrian Lahoud writes, with reference to the question of structural violence, the fear or accusation that distribution equals suspension proceeds from a false logic of "all or nothing" (which, we might add, can be quite knowingly designed to keep certain zones of responsibility away from scrutiny). Against this, the challenge is "to discern with enough precision and attention the action of this network and its agents, to follow the threads of relation where they lead, to rigorously unfold the complex causal links" (2014: 514). This done, what emerges is not an indistinct agential amalgam but a clearer, better-differentiated map of the kinds of activity at work at different points of the alliance, and the kinds of responsibility that these sites might accordingly be thought to bear.

In the case of the left-to-die boat, we have a scenario that poses precisely this question of just how a model of distributed agency accounts for the responsibility of different participants, inasmuch as the combative alliance that emerges around the work of Forensic Architecture includes the respective contributions of human and nonhuman participants, notably the survivors. I would argue that if survivor testimony features alongside that of maritime charts, satellite images, military logs, and so on, this produces not an offensively indifferent leveling but, on the contrary, a sharply delineated picture of the different participants' roles. And these participants, moreover, are defined not by the categorical assignment of essential qualities but by participation in their respective alliance: it is *as participants* that they feature, each in their different way, in a joint venture of confrontation. Survivors here become witnesses; that is, when they figure in alliance with the sea, its analysts, desertification, poverty, and so on against the military, legal, and governmental forces that themselves feature only as participants in the alliance acting to secure this deadly mode of maritime governance. Participants contribute differently; the action of confrontation is the work of the alliance. And this confrontation is decisive: it is through this that the nature and significance of the events in question are determined.

At this point, we can start to reflect on the conception of responsibility at work in this model of decisive, distributed agency. As we have seen, the legal context of the majority of their work makes the notion of responsibility loom large in Forensic Architecture's conceptual landscape. And as we have also seen, this is a context in which meaningful responsibility is not attributed solely to sovereign individual human beings. This expanded attribution is less unusual than it might appear: beyond the familiar legal concept of corporate responsibility, notions of distributed responsibility in fact abound in everyday scenarios. To give only three examples: sports teams, families, and criminal gangs are routinely described as exhibiting an emergent responsibility that is not reducible to that borne or exercised by any of their individual members. (For a comprehensive survey of relevant theories of "group agency" here, from the legal to the everyday to the philosophical, see Tollefsen 2015.) But as Floridi and Sanders point out, most existing accounts of distributed responsibility are prepared to extend the concept in this way only insofar as the agents involved are either human individuals or "human-based"; that is, "reducible to an identifiable aggregation of human beings" (2004: 350). Floridi and Sanders make this observation in the context of theorizing what their title calls "the morality of artificial agents"; and, of course, the accelerated development of forms of artificial intelligence has brought the question of nonhuman agential responsibility dramatically to the fore. We will look at the specific questions this poses in horizon 2. For now, though, we should consider a little further the understanding of responsibility at work in the practice of Forensic Architecture.

I will note, first, that just as throughout this study I will not attend greatly to distinctions made elsewhere between "actor" and "agent," given that these mostly serve to secure a priori distinctions between human and nonhuman participants (see the introduction), so will I move freely between "accountability" and "responsibility," with no concern to reserve the latter (or a "strong" version thereof) for some given type of being. The work of Forensic Architecture has already allowed us to see an internally differentiated conception of responsibility at work in the analysis of distributed, conjoint human-nonhuman agential alliances, without this differentiation being indexed to participants' supposed a priori properties. Clearly, this conception has dispensed with reflexive consciousness as a necessary condition for the attribution of meaningful responsibility: the Kantian "I think" is plainly not a factor in the representations produced by satellites, let alone the tides of the sea. But as I have been arguing, there is no need here for human-exceptionalist alarm. In the first place, as we have seen, responsibility is meaningfully attributed every day to composite agents without these agents being understood as a unified reflexive consciousness. In the second place, and more fundamentally, understanding responsibility as emergent along with agency along a composite agential front produces not a smearing of responsibility indifferently across all participants but *more and better differentiation* of the distribution of degrees and kinds of responsibility through their alliance, thanks to which some participants might be identified as what I am calling "eminent sites of accountability." Such differentiation will necessarily entail reference to capacities; not as inherent properties of given kinds of being, however, but as themselves emergent and differentially distributed, relative to a participant's position in the alliance.

This approach to accountability is already imposed, for example, by seeing the emergence of an alliance as an "agential cut," in Barad's term (1997: 140): decisively reconfiguring the elements of its situation, it is this emergence that transforms beings into participants *in the act*, defined not by a priori attributes but by their relative positions in and contributions to the alliance. As has indeed been the case here: the sea, a satellite, and a ship's captain are accountable in different ways according to the capacities with which they emerge as participants in the alliance and can duly be held differently responsible. And note: analysis of the captain's responsibility will be all the more precise if this is differentiated from that of the other participants, in relation to which this responsibility is constituted. As Bennett writes, "Human intentionality can emerge as agentic only by way of such a distribution" (2010: 36). In its full form, such differentiation will also entail reference to futural projection and commitment, in some cases even in the absence of reflexive consciousness; but we are not yet in a position to elaborate this full form, for which we will require the following chapters. What we have seen so far, though, is an engaged analytical practice that makes effective use of a notion of distributed agency as decisive and as compatible with the meaningful attribution of responsibility.

Given the nature of this practice, it is not only understandable but inevitable that the researchers at Forensic Architecture should frame questions of responsibility—and, indeed, account for their activities as a whole—in the terms of legal representation and debate. Although it might seem to embroil them inescapably in the anthropocentric proceduralism we saw compromise Latour's expansive parliamentarianism—or, rather, precisely because it does so—this idiom has the signal advantage of bringing this differentiation of responsibility unmissably to the fore. "Things do not speak by themselves," writes Weizman, "they have to be asked" (2016: 104). If forensics makes things speak, there is, of course, no escaping the specter of ventriloquism, the puppet master putting words in the puppet's mouth. But this specter is entirely welcome here, for two reasons. First, manipulation is one pole on the spectrum of allegiance, as conscription characterized by a maximum of compulsion. Its shadow presence here reminds us that conscription is always an encounter between unequal forces, even when this is not in so extreme a mode as manipulation. Second, the inequality this discloses calls for delineations of responsibility that will be ever sharper, ever more carefully differentiated.

"Making speak" can be abusive; it can also be enabling. The same is true of "speaking-for." Where the Latourian model of delegation can get stuck in the inconsistencies of its all-too-human attachment to speech, the version we find in the work of Forensic Architecture fares rather better, letting its expert delegates be more or less human, more or less technological, but always insisting on their delegation as situated somewhere along the spectrum of conscription, from friendly advocate to rough manipulator. As in the description of the sea being "interrogated and cross-examined as a witness," there is here no idealistic effacement of respective interests and positions of power: "*we brought the sea to bear witness* to the conditions that have led it to kill," write the researchers in question (Heller and Pezzani 2014: 674, 673; my emphasis). There is, however, an understanding of such interests and positions as constituted by their relations to other positions within the relevant alliance. As we have seen, the interventions of the forensically informed advocate or activist prove effective—that is, manage to "speak," to make representation within a forum—only when they can ally themselves with the evidence of human and nonhuman witnesses. As Bennett writes, "insofar as anything 'acts' at all, it has already entered an agentic assemblage" (2010: 121). If some participants emerge as inflecting the interventions of the alliance more forcefully than others, these respective capacities are constituted by their differential distribution among those who now feature as participants. And it is the alliance that acts.

FURTHER CONFRONTATION

Throughout the work of Forensic Architecture, then, we find a model of distributed agency that is eminently political: in which combative alliances of diverse participants decisively confront each other in ways that dramatically engage the lines of force shaping their differential composition. This work has accordingly allowed us to envision how antagonistic alliances confront one another. It might also perhaps reassure Latour, in its demonstration that exemplary attention to the contributions of heterogeneous participants is wholly compatible with a combative orientation; or, to put it another way, that a combative orientation can be effectively maintained without recourse to the destructive human-exceptionalist fantasies of ontological transcendence. But if Forensic Architecture has thus allowed us to develop a model of the decisively confrontational alliance, we now need to look more closely at the organization of such confrontation, for the alignments within and across alliances are still more complicated than the schema we have established so far might suggest.

We will begin this consideration with an example familiar from chapter 1: Latour's claim that the Anthropocene divides the inhabitants of the Earth into two nested pairs: human beings versus the rest, and, within the human beings, Humans versus Earthlings. As Déborah Danowski and Eduardo Viveiros de Castro point out, however, if agency indeed operates at the level of alliances of human and nonhuman actors, then the division between Humans and Earthlings "is not only internal to our species" but "opposes two camps or sides populated by humans and nonhumans—micro-organisms, animals, plants, machines, rivers, glaciers, oceans, chemical elements, and compounds. In short, the whole range of existents that find themselves implicated in the advent of the Anthropocene" (2017: 100–01). With the battle lines configured in this way, debates over species-level responsibility for climate change, say, can look like so many local skirmishes. For in those debates, there is no question that human agency is the issue; the only questions are whether this agency is that of all humans taken as a species (and, if so, how this should be exercised) or only some. (For these respective positions, see Chakrabarty 2009; and—from among the many responses to this paper—Malm 2015; and Keucheyan 2016.) By contrast, Jason W. Moore's theorization of what he terms the "Capitalocene" manages both to sustain an incisive critique of capitalism as ecologically devastating and to conceptualize agency along the lines of the antagonistic human-nonhuman alliances sketched by Danowski and Viveiros de Castro. "Relations of class, capital, and empire are *already* bundled with extra-human natures," writes Moore: "they are configurations of human and extra-human natures. From this it follows that *agency is a relational property* of specific bundles of human and extra-human nature" (2015: 37).

Building on Moore and Danowski and Viveiros de Castro, I want now to develop my claim that strong confrontation is compatible with distributed agency. Division divides, fractally and fractiously: not just humans against humans or, indeed, Humans against Terrans, or humans against the rest, but composite against composite, alliance against alliance, on each side.

As I will now argue, this field of conflict is dynamically organized by the relations of force obtaining within and between composite alliances, relations of conscription, instrumentalization, fidelity, infection, solidarity, existential negation: of imbrication and fully frontal antagonism. In what Nigel Clark and Kathryn Yusoff call "the geologization of the political" (2017: 17; adapting Deleuze and Guattari's quasi-Nietzschean "geology of morals" [2014: 45]), action is conjoint but no less conflictual, and as we saw earlier, hierarchies of scale (or of part and whole) can implode. Earth and ocean, trawler net, whale, and microbead, rainforest, logging company, government, and biofuel pursue their struggles across a field traversed by combative alignments and within which, for that very reason, power differentials multiply. Duly qualified, Latour proves helpful here: describing the new *geo*-politics as "the *addition* of powerful forces of resistance thrust into the class conflicts and capable of transforming their stakes," he affirms that "the outcome of the disputes can only be modified if all rebels, in overlapping configurations, are entrusted with the task of fighting" (2018: 61). As a result, "we are going to be able to multiply the sources of revolt against injustice, and, consequently, to *increase* considerably the gamut of potential allies in the struggles to come." "From now on, we benefit, so to speak, from help offered by unleashed agents" (88). If we subtract what we now know to be Latour's habitual displacement of the conflict (here, from the present to the near future), we are left with a very useful model, in which the agency of the various combative alliances can be understood as the joint mobilization of the emergent capacities of all manner of "unleashed agents."

In his *Carbon Democracy*, Timothy Mitchell provides an outstanding example of what such a scenario might look like in his interpretation of the connections between fossil fuels and the forms of political power and geopolitical conflict that have defined the industrial era. Mostly, Mitchell follows a classically Latourian line, emphasizing that actors we might think of as drawing up and implementing plans for rapacious domination have usually, in fact, been improvising strategies to shore up their vulnerable position. (Latour returns the compliment by drawing on Mitchell in *Down to Earth*; see Latour 2018: 62.) At one point, however, Mitchell goes beyond this emphasis on the weakness of the apparently powerful and shows how confrontational political intervention can emerge from particular configurations of human and nonhuman actors.

Discussing the successes of the labor movement in North America and Europe between the late nineteenth and mid-twentieth centuries, he analyzes the contribution made by the dominant methods for extracting and distributing coal to the vanguardist power of mining unions. Vulnerability is again key—but in this case, it is the vulnerability of industrial processes as a whole, created by the combination of their dependence on coal and the infrastructure of its distribution via concentrated channels that are susceptible to interruption by intervention at a single well-chosen point. To this, Mitchell adds the topography of labor organization: working underground and among themselves, miners were able not only to build solidarity but, crucially, to organize away from the surveillance of their employers. Mitchell thus understands the power accrued by mining unions in this period as an emergent property, "a new form of collective capability built out of coalmines, railways, power stations, and their operators" (2013: 27). Each element of which is of course susceptible to further parsing: for "coalmines," say, we need also to understand coal, wood, capital, steel, muscle (human and equine, at least where horses, donkeys, and ponies replace or accompany women and children as mineworkers), not to mention the overwhelmingly female labor of social reproduction by which the work of the mine is sustained. Across the alliance, capacities are relationally constituted and differentially distributed and valorized. Alliances clash and combine; and when it combines with others that amplify its power (the wider labor movement and its channels of communication, public transport, etc.), from the alliance we call "coal mining" there grows the interruptive force of the general strike, the quintessential image of confrontational workers' mobilization landing a decisive blow in the class struggle (Mitchell 2013: 23–25).

Where Latour's tendency to polarize the political field into the false alternative of discussions versus bloody Terror appeared to deny this possibility, then, we can now see that a frontally combative intervention such as a general strike can indeed be meaningfully understood in terms of distributed agency without this combative quality being frittered away in a bad infinity of ineffectivity. We should remember, though, that such agency is not necessarily on one side or the other of any given confrontation. Mitchell contrasts the affordances of the mining industry for labor power with the governmental politics and distribution infrastructure of the oil industry, which he reads as inimical to workers' power and as maximizing companies' control over the rate and continuity of supply. We can also contrast his description of these affordances with the account given by Andreas Malm (2016) of the rise of fossil fuel capitalism in the conversion by their owners of British textile mills to coal-fired steam power, cheaper and more reliable than water power and human workers. In these examples (and reading Malm against his human-exceptionalist commitments), we can see how agential power on the other side of the class struggle from Mitchell's miners is also composed of human and nonhuman capacities, here including those of mineral and topographical participants. And through this, we see again the point made by Danowski and Viveiros de Castro: that a given alliance can appear on both sides of a given conflict. The alliance we call coal mining is recruited by millowners to reduce their costs and secure consistent productivity. Having become essential in this and many subsequent ways, it can also launch crucial interventions by organized labor. As in a conventional Marxist understanding of the affordances of industrialization for workers' organization, the same alliance finds its emancipatory force within the structure of its current exploitative mobilization.

Thanks to Mitchell, we can thus understand even a large-scale class confrontation such as the general strike—and, indeed, the forces of the state and of capital that seek to suppress such interventions—as the work of antagonistic alliances composed of participants of all kinds. Bringing my notion of such antagonistic alliances into proximity with the language and conceptual framework of labor militancy in this way also provides an opportunity to clarify the precise differences between this notion and William E. Connolly's advocacy, in *Facing the Planetary*, of what he calls "militant citizen alliances," which might serve to "mobilize cross-regional general strikes" (2017: 9, 129)—and through this, between our respective approaches more generally. In *The Fragility of Things*, Connolly called for "a more militant democratic politics," embodied in "large-scale, cross-state citizen actions," "alliances between minorities of multiple types who join together to reorient the common life" (2013: 10, 42, 137). In this subsequent work, he builds on this, seeking "an idea of the general strike appropriate to the contemporary era" (2017: 129): namely, one that would "explicitly factor in the real power of a variety of nonhuman, partially self-organizing systems entangled with market and other cultural processes" (137). Clearly, this idea offers many points of comparison to my attempt to understand political confrontation as the work of heterogeneous antagonistic alliances. Since its publication, moreover, this formulation has proved extraordinarily prescient, with the advent of regular climate strikes "that are both general and cross world regional lines" and that mobilize "a militant, cross-regional pluralist assemblage" (144).

This remarkable power of anticipation speaks volumes for the analytical force of Connolly's attention to the composite nature of our defining geopolitical processes. Moreover, the development of his conception of militancy in the direction of the outright confrontation that is the general strike firmly underlines the compatibility I am arguing for between a model of distributed agency and the possibility of decisive, at times confrontational political intervention. There is, however, a significant difference between our approaches. This is, I think, a matter of emphasis or perspective; and I certainly envision these approaches as working in solidarity, pushing in the same direction. This is not to say, however, that what separates us is trivial. For as his reference to "militant citizen alliances" suggests, Connolly's actors are human: his "evolving, complex, cross-regional 'we' that propels diverse constituencies into larger assemblages, even as each constituency retains a host of differences" (2017: 34) is diverse in terms of its *human* participants. Of course, this must be qualified at once: in Connolly's understanding (see especially Connolly 2010), the actions of these human participants are unthinkable other than as emergent from the pre- or proto-agential activity of all the nonhuman participants and processes that create the possibilities these actions might realize, and the boundary between human and nonhuman contributions is here maximally porous.

Nevertheless, Connolly's aim is to encourage better-informed, more ecologically attentive forms of effective human action. His invocation in *Facing the Planetary* of the global general strike is an inspiring part of this project; but this is where our emphasis, or our perspective, differs. Connolly's militant action is that of human beings, who understand themselves as acting on behalf of and thanks to a broad collective of diverse beings. Mine is that of this collective, whose different participants intervene in different ways. These interventions are not equivalent to each other; they culminate in the action of the alliance, but none of them gathers or represents the others. The "multiple types" who here join together "to reorient the common life"—including, on one side, in the direction of more extreme and more capillary exploitation—participate alongside each other in the action of a broad alliance. Some are nested within others; some conscript others; and some are certainly more powerful than others, generally thanks to such conscription. But from irreducible contributors to forms of action that finally remain human, all are now considered participants, and the action is that of their alliance.

As I argued in the introduction, in my view, the ambition to account for political agency in terms of distributed agency—as opposed to motivating better forms of human action—makes such a shift of emphasis necessary. It is only a shift of emphasis—but it is not nothing. For the site and nature of action shift with it. Consider the following, from the Invisible Committee: "The world doesn't environ us, it passes through us. What we inhabit inhabits us. What surrounds us constitutes us" (2015: 79). Both Bennett and Connolly would happily agree with this, I think; but their understanding of "passes through" would feature "us" as the point at which the force in question is focused (and rightly so; they are interested in the quality of human action), whereas mine would see this force passing through and outside "us," through and outside all kinds of others, differently inflected in each case but continuous and making its impact through all these, in their dispersed composition. I want to think of the agential alliance as a *front*, in an image forged from meteorology, politics, and conflict together: an extended, mobile alignment of participants whose force—combative where necessary—is registered cumulatively right along its length, while at its local positions participants advance and retreat according to the prevailing possibilities.

Modeled as the activity of antagonistic alliances, then, decisive action emerges from a whole composite of human and nonhuman elements without necessarily being focused through any one of these and without thereby losing anything of that decisive character. On the contrary, it is precisely thanks to this composition that the action emerges as decisive, as it brings into play as the stakes of its intervention the full set of vital relations that give meaning to the existence of the array of beings making up its situation (see Bineau and Chopot 2017: 49–51). And this is why we find ourselves somewhere between the poles of mere efficient causality, on the one hand, and agential decision-making, on the other. Not only because this distinction cannot survive any sustained attention to the multiple factors constituting such decision-making (although, indeed, it cannot, as we will see throughout the following chapters), nor because some of the processes involved in such emergent action display "teleo-dynamic" qualities with a more complex relation to finality than is captured by their confinement within the bounds of "efficient causality" (although they certainly do; and on the "teleo-dynamic," see Connolly 2013: 8). But also, and especially, because the presence in this way of multiple, differently interested parties brings a thickly motivational dimension to the various actions of the alliance all along its front that the mechanistic character of "mere" causality can only miss. For this reason, the first sense in which I am here understanding "decisive" action—namely, as making a major difference to a situation—cannot simply be dismissed as "utterly minimal" and "hard to tell from the property of existing," as Malm writes of Latour's similar definition of agency (as "making some difference to a state of affairs" [Malm 2018: 89, 81; citing Latour 2005: 52]). The human-exceptionalist polarization of the scene of action into mechanistic causality and human decision-making evacuates the action in question of the very dimension that makes it decisive: as Latour would argue, without a sense of the pressing claims of all participants, how can anything like an adequate account of the significance of this action be produced? The sovereign human decision, exercised with however detailed an appreciation of the relevant "causal" factors, will always miss just what gives this situation its urgency, and so what makes the action in question decisive in this first sense.

If I am trying to think the decisive actions of antagonistic alliances as differentially distributed right along their front, then, it is to forestall even the minimal priority that comes with understanding human action as the focalizer of its allies' interests, to resist that gathering for as long as possible and so to see the force of these interests as itself shaping the decisiveness of the actions in question. I am concerned that attaching even a relative privilege to human action will compel us to enact such solidarity at best in the mode of stewardship, and that this mode can struggle to avoid drifting into instrumentalization—as we see in certain prominent geoengineering "solutions" to the climate crisis, with their unshaken commitment to the extractive economics and domineering metaphysics that have generated the crisis they seek to avert. As I argued in the introduction, my sense is that—despite what I agree is the sheer difficulty of doing this, given the inescapable existential situatedness of all beings—placing human interests rigorously *alongside* those of all participants as the factors determining an action as decisive can offer a helpful way to resist the pull of that instrumentalization. I thus want to add my approach to that of existing theorists of distributed agency, to contribute to the inspiring work they are already doing as part of this resistance, and to address an area in which I feel this work needs further reinforcement, by suspending still further any reference to specifically human action as the domain in which concerns are meaningfully registered or commitment focally concentrated.

As Latour pithily puts it, then, if red tuna are dying out, this is already a pretty clear form of self-expression (Latour 2012b: 146). Less succinctly, we might consider the floodwaters that increasingly visit European urban areas. These already describe the damaging consequences of covering ever more land with asphalt without regard for local ecosystemic cycles of rainfall absorption; they did not have to wait for me to write this sentence in order to make this point. In seconding that point here, I am bringing it into representation in this forum, "making them speak" in this idiom, conscripting their strength for these local purposes; but I am not, I think, translating their actions into a higher order of agency. I am not acting on their behalf, or on behalf of the various inhabitants of these ecosystems, but am joining their front in common cause—and the cause has mobilizing force precisely because it enacts a "struggle to define the modalities of life in common" (Marchetti 2013: 14) in which the relevant commonality joins together all manner of beings, from the very beginning, and crystallizes the stakes of its milieu. The action in question—here, confrontational denunciation of an appropriative mode of inhabiting the land—is that of the alliance as a whole (floodwaters, critical urbanists, ruptured sewers, displaced wildfowl, concerned citizens, hungry rodents, theorists . . .); its stakes are given by their collective interests; its force is registered differentially at the positions occupied by its participants (flooded basement, river mouth, town hall, meadow, conference hall . . .); and it engages *frontally* the alliance of those whose interests it opposes (water companies, other rodents, other wildfowl, city planners, Salmonella typhi . . .). The burning rainforest, the flooded underpass, the raging bushfire declare emergencies that go beyond human priorities, *including* the priority to attend better to their declarations, and it is because of this expansive existential commonality that they constitute decisive events. Mobilizing concerns that can be understood only as overspilling the interests of any one participant, they bring into play alliances—for example, government, speculative finance, timber exporters, construction companies versus indigenous peoples, the carbon cycle, species and ecosystems unknown to scientific classification, eco-activists—whose respective claims compel action in the mode of confrontation.

This is, perhaps, a first image in our study so far of what politics looks like, once we replace human exceptionalism with a version of distributed agency that can encompass decisive, conflictual intervention. The town hall is still there and still full of human beings giving voice to their concerns. But now we can understand these concerns as joining and clashing with those of all kinds of other beings and acquiring their urgency—as Latour would insist—precisely from their position within this mesh. What is more, we can understand those human participants—the active citizens of our political traditions—as local sites along the agential front formed by the actions of an alliance, with their particular capacities emerging in the differential distribution of capacities across the alliance, and their activism—its intensity and effectivity—defined by their participation in this alliance. And, finally, we can understand the decisive clash that happens in the town hall, at the protest camp, or at the pipeline as the confrontation between antagonistic alliances and politically decisive precisely because of the drastically expanded stakes crystallized in their actions.

* * *

To be clear, the claim that distributed agency is compatible with decisive, confrontational action is not yet a theory of the decision. If we have established that such agency can be decisive in the first of our senses—as making a major difference to a state of affairs, beyond mere efficient causality—we have yet to see it as decisive in the second of these senses; namely, as taking a strong position, engaging a partisan commitment. In common with most approaches to distributed agency, our perspective thus far has remained essentially retrospective: to incorporate an account of the decision in the strong sense, we will have to face the challenge of understanding futurity without transcendent intentionality. This will require a model of the decision and of futural commitment as processually immanent, which is what Stiegler and Malabou will help me to develop in the following chapters.

But let me again be clear: even once this is in place, I will not be arguing that political decisions are made just as well by train tracks and streams as they are by miners and millowners. Once again, the point is not a morass of undifferentiated agency but more and better differentiation. I am, then, certainly arguing that what we think of as the punctual political decision emerges as a decisive intervention thanks to the relations obtaining within its alliance; that it is thanks to these relations, and those of the alliance to its wider situation, that it gains its decisive character; and that the intervention is that of the alliance as a whole, its impact distributed differentially all the way along its front, not focused at the position of one participant.

Allow me accordingly to repeat and unpack this core claim: the ontological egalitarianism of distributed agency demands not the abolition of differentiation but the emergence through a specific alliance of multiple differentially distributed and interreliant capacities, whose distribution among diverse participants is not given but, rather, as an index of relations of domination and emancipation, shifts across time and space to delineate at any given moment the composition of this expanded political field. As Phillip John Usher shows (2019: 70), parsing by species is a way to miss the differentials making up the politics of a given field; by contrast, intensifying our understanding of the agential alliance at work brings this politics into more realistic and, perhaps, more effective focus. How *does* the sea become both killer and grave? How *do* survivors become witnesses? How *do* the responsibilities of multinational petrochemical corporations for the climate crisis differ from those of the various beings inhabiting coastal regions—and how are these differences elided? In all cases, asking these questions means understanding that it is the alliance that acts and that as a consequence, responsibility is no less composite than agency. Once this is accepted, action and responsibility can meaningfully be parsed into their constituent contributions and multiple sites, and their emergence and differential distribution can be more sharply delineated.

To return to the ecological activists with whom we started this section, then, we can indeed understand human climate militants as acting with or as a continuum of nonhuman and human beings—but we can do so because there is no such action otherwise. Our ecological activists join with factory owners, pit ponies, survivor witnesses, weather patterns, surveillance satellites, and the sea as participants in alliances joined in decisive confrontation. The Anthropocene did not invent distributed agency—although it has made the denial of its reality less credible and has laid down the task of conceptualizing militant activism along these lines. As Clark and Yusoff write, then, "relevant collective action must be understood as being not only about or towards the earth but emerging with or through the earth" (2017: 17). Or, as Latour puts it: "class struggles depend on a *geo-logic*" (2018: 62). As Latour never quite manages to say, however: the conflict in question is already here.

2

BERNARD STIEGLER

Deciding on the Accident

The model of antagonistic alliances aligned in decisive confrontation, as developed in the previous section, is designed to supplement the limitations of the Latourian account of distributed agency when it comes to the question of conflict—and hence to the question of politics. As mentioned, however, to address this question, our model will also have to offer a futural dimension; only then will it allow for the partisan commitment and prescription I am here taking as indispensable elements of any adequate account of political activity. As developed so far, this model remains essentially retrospective: the decisive quality of an intervention is not given in advance but established through the intervention itself. At this point, then, we need to integrate the second sense of the *decisive* intervention: as not just making a drastic difference to a situation but also proceeding from a decision. Accordingly, our task now is to formulate a theory of the decision as an act of commitment and prescription—a projection—that will prove compatible with the egalitarianism that motivates theories of distributed agency by refusing to secure itself through human-exceptionalist appeals to the ontological transcendence offered by the capacity for such projection in time, and instead remaining immanent to the process of its emergence. It is time to open our model to the future.

In search of this opening, I now turn to the work of Bernard Stiegler. If Stiegler is best known as a philosopher of technics, a critic of the libidinal economy of capitalism in the era of neoliberalism, or a proponent of forms of social organization that might address the planetary crisis labeled the Anthropocene, these aspects of his work will not be my main concern. They will feature, of course; in some cases, quite prominently. But what I will take from Stiegler is not so much this or that position or theory; what I will take from him is a structure. And this structure is that of the processually immanent decision as an interventionist moment of futural commitment, which emerges in the wake of an accident: the projection and prescription of a future in the mode of reflexivity without transcendence.

Stiegler builds this structure from two key parts of his thinking. First, combining Simondon's concept of individuation with Deleuze's notion of the quasi-cause, he is able to describe the emergence of a locus of commitment without recourse to preexistent, categorically given capacities. Second, via Leroi-Gourhan's account of the coconstitution of human beings and their tools, he specifies that commitment as operative primarily in decisions as to the adoption of particular technical forms—and, moreover, offers such decisions as a definition of politics. (We will work though all of this in detail later.) As well as delineating this structure, I will also accordingly consider Stiegler's discussions of particular contemporary political questions in order to develop a sense of what this politics might look like in practice.

Stiegler's idiom differs considerably from the one I have been using in this study so far. His key terms are *individuation*, *adoption*, *epokhè* (and, latterly, *entropy*, *negentropy*, and *exosomatization*); if he occasionally uses *agent*, and if he is a major thinker of the *accident*, he certainly does not think in terms of *agential alliances*. In a sense, of course, Stiegler's understanding of the coconstitution of the form of life we call human and its originary technical prostheses does offer a kind of agential alliance in which what acts is the composite of these two. But for him, this is not itself a major concern. One consequence is that in this chapter, we will need to spend some time inhabiting Stiegler's lexicon in order to get to the structures I am looking to adopt from his work. The goal of the chapter is not to unpack Stiegler's philosophy; although I will do a fair amount of such unpacking, my main interest is to identify what Stiegler's thought offers to the model of agency I am building throughout this study. I will borrow from Stiegler key conceptual resources for this model, especially as regards the *decision*; I will not, on the whole, adopt much of his idiom. But Stiegler's contribution will be essential, for it is thanks to him that we will be able to understand the decision—in the strong sense of the term—as both processually immanent and engaging a normative—indeed, partisan—commitment.

We will proceed as follows. An initial section on Stiegler's elaboration of his core philosophy through readings of Leroi-Gourhan and Simondon will be followed by an account of his model of the decision as both processually immanent and strongly interruptive. The operation of this model will then be traced in his discussion of three particular questions: the "industrialization of memory," the politics of automation, and the planetary crisis indicated by the notion of the Anthropocene. A final section will then sum up exactly what I am going to be able to take from Stiegler, where I part company from him, and what still remains to be done.

INHERITANCE

For our purposes, then, the key aspect of Stiegler's work will be his formulation of the decision as a moment of reflexivity without transcendence. In order to understand this formulation, though, we need to work our way through some of the building blocks of his philosophy, starting with the essential role played by his reading of two essential figures: philosopher of technics Gilbert Simondon and, first, paleoanthropologist André Leroi-Gourhan.

Stiegler initially comes to Leroi-Gourhan via Derrida, whose *Of Grammatology* discusses Leroi-Gourhan's *Gesture and Speech* in some detail. (See Derrida 1976:74–93; and on Derrida as Stiegler's gateway to Leroi-Gourhan, see Haworth 2016:153–54.) In his own reading, Stiegler emphasizes especially that passage of human evolution initiated by the appearance of the hominid formerly known as *Zinjanthropus boisei* (later renamed *Paranthropus boisei*), about 2.4 to 1.4 million years ago (see Stiegler 1998:134–79). Discovered by the Leakeys in 1959, *Zinjanthropus* possessed a small brain though already using stone tools. What this shows for Leroi-Gourhan is that tool use is not the consequence of brain expansion, as had been assumed. Brain expansion instead follows a change in posture, as the upright stance frees the hand for the tool use that will accompany this expansion. What is important here for Stiegler is this entwined development of cortex and tool: as he puts it, over the course of this period, "cortex and equipment are differentiated *together, in one and the same movement*"; "differentiation of the cortex is determined by the tool just as much as that of the tool by the cortex" (1998:158). This gives Stiegler his key point: that in a "singular process of structural coupling" (158), the form of life we call "human" is coconstituted with its technical prostheses. (For a carefully argued critique of Stiegler's reading of Leroi-Gourhan, incidentally, see Johnson 2013.)

The notion of tool use as consequent on brain development imagined such tool use as the exteriorization of a prior capacity "internal" to the human. The coconstitution Stiegler finds in Leroi-Gourhan, by contrast, rules out this essentially metaphysical model, with its attachment to a reflexive "interiority" supposedly proper to the human. As Stiegler writes, "The paradox is to have to speak of an exteriorization without a preceding interior: the interior is constituted in exteriorization" (1998:141). Instead of interiority and exteriority, we should think of "an originary complex in which the two terms, far from being opposed, compose with one another" (152). As Stiegler summarizes in the second volume of *Technics and Time*, this is "the process of corticalization taking place simultaneously with the tool's appearance—creating a new rapport between living being and environment, mediated through an artificial layer qua artificial memory as essential to human being" (Stiegler 2009a:164). For this reason, Stiegler describes his project in the first volume of *Technics and Time* as an "archaeology of reflexivity" (1998:140): the reflexive interiority supposedly proper to the human not only has a material history whose traces are studied by the archaeologist; it exists paradoxically *in its relation to* and composition with its "outside." (On this, see Lewis 2013.)

To specify the structure of this originary complex, Stiegler mobilizes an understanding of the *prosthetic* developed by Jacques Derrida in his analysis of the operation of the supplement: namely, the strange insight that the prosthesis somehow completes something already defined as whole. (See Derrida 1976:141–64 and 313–16.) The technical—which is not the human—is integral to the constitution of the human: "With the advent of exteriorization, the body of the living individual is no longer only a body: it can only function with its tools" (1998:148). The human is in this sense a distributed system, paradoxically composed of itself *plus* its technical prostheses: "The prosthesis is not a mere extension of the human body; it is the constitution of this body *qua* 'human' (the quotation marks belong to the constitution)" (152–53).

As Stiegler presents it, the appearance of *Zinjanthropus* puts an end to the genetic evolution of the human: "the most archaic technical evolution is already no longer 'genetically programmed' " (1998:135). After the subsequent stabilization of cortical organization, human development proceeds exclusively through the development of technical forms, meaning that "the evolution of this essentially technical being that the human is exceeds the biological." "The evolution of the 'prosthesis,' not itself living, by which the human is nonetheless defined as a living being, constitutes the reality of the human's evolution, as if, with it, the history of life were to continue by means other than life" (150). Most notably, the inscription of traces of individual and collective existence (including accumulated knowledge) in what Stiegler calls "organized inorganic matter" (49) means that the memories of the group and the individual can be preserved beyond death and inherited *as a past they have never lived* by subsequent descendants. Contrasted with "a 'program' in the *quasi-deterministic* biological sense," this is what Stiegler terms "epiphylogenesis," which he defines as "the conservation, accumulation, and sedimentation of successive epigeneses, mutually articulated," and which, he emphasizes, represents "a break with pure life, in that in the latter, epigenesis is precisely what is not conserved" (140). Human evolution is epiphylogenetic and takes place as the transmission and development of technical prostheses.

For Stiegler, " 'human nature' consists only in its technicity, in its denaturalization" (1998:148), and "The unity of the human . . . becomes tenuous: one can hardly see any other permanence . . . than the fact of technicity" (149). For this reason, Stiegler is happy to identify technical epiphylogenesis as more than human. In the second volume of *Technics and Time*, for example, he writes, "it is clear that processes of the transmission of experience and 'acculturation' occur in colonies of great apes, and it seems more and more obvious that the process I am calling *epiphylogenesis* here originated well before the advent of the human—which confirms that the issue is not that of 'the human' but of the process of which it serves as the transmitter" (2009a:255n44; translation modified). If the human certainly cannot be thought without the technical, the two are not therefore the same; nor is the human essentially and exclusively *the* technical being. Rather, Stiegler's reading of Leroi-Gourhan allows him to argue that within the epiphylogenetic development of the human, human beings and technical forms are coconstitutive, entwined in such a way that it is in fact the relation that constitutes its terms. In the language of Gilbert Simondon, such a relation is termed *transductive*; and the transductive relation between human beings and their technical forms is precisely what Stiegler will take from his reading of Simondon, modifying the latter's thought in order to do so.

In his critical elaboration of Simondon, Stiegler moves from considering the coconstitution of the human and the technical as this is evidenced in paleoanthropology, to the processes of what Simondon calls "individuation," the emergence of individuated beings from what he terms the "preindividual milieu." According to Stiegler, Simondon's "absolutely original conceptual framework" describes what Stiegler summarizes as "processes of vital, physical, and psychosocial individuation"; namely, the processes giving rise, respectively, to the emergence of individual living, nonliving, thinking, and social beings (2009b:66). From this range, Stiegler focuses especially on the relation between the last two, highlighting their irreducible imbrication. In Simondon, he writes, "individuation is conceived as a *process* which is always *both psychic and collective*—where *I* and *we* are therefore two aspects of the same process, and where the *difference* between them is also the *dynamic* of the process" (2014:45). Individuation, he writes, "is always already at the same time psychic *and* collective"; just like the coconstitution of human beings and technical forms, moreover, "this individuation surpasses the opposition between interior and exterior" (2009b:65).

Drawing on the point emphasized by Simondon in the opening pages of his major thesis, *L'Individuation à la lumière des notions de forme et d'information* (Simondon 2013), Stiegler stresses that the starting point here is not the already constituted individual but, rather, the processes of individuation that allow an individuated being to emerge from the preindividual milieu. Following Simondon, Stiegler argues that "the preindividual must now be thought of as being the environment in which a process occurs whose result is the individual, and for which the categories that allowed us to discern the individual are inoperative" (2015c:66). "The individuation in which I participate in this way is not therefore mine alone" (2014:47): "psychic" individuation is situated within "a larger collective individuation," in which process, potentials from the preindividual milieu "are actualized by psychic and social individuals" (2015c:70). There is a clear resemblance here to the structure of epiphylogenesis: just as in that process, later generations inherit a memory that they have not lived; so in the process of psychic individuation do "I" come into being by virtue of my relation to a milieu and a collective that exceed me. That is, psychic beings are always already *psychosocial* beings. (For an excellent discussion of Stiegler's relation to Simondon, including an account of the beleaguered publication history of Simondon's writings, see Barker 2013.)

Stiegler highlights in particular Simondon's use of the concept of metastability to describe these operations: the metastable preindividual milieu gives rise to the emergence of stabilized individuated beings, just as crystals form in a supersaturated solution. (Although Simondon's description of crystallization is an extremely helpful image of the broad outlines of the process of individuation, to the point that he himself occasionally uses this analogy [e.g., 2013:296], it should nonetheless be noted that he distinguishes carefully between the details of this process as it gives rise, respectively, to "physical" and "biological" individuals [see Simondon 2013:157–62].) "The preindividual," Stiegler writes, "is only individuated through the stages of a process of individuation in which one metastable state follows another, always on the verge of a disequilibrium" (2015c:70; translation modified). Bearing within itself unactualized traces of the preindividual milieu, "potentials . . . for an individuation to come," "the individual finds itself always already surpassed by the individuation that traverses and supports it" (71). Individuated beings can (and do) always deliquesce, lose individuation, and return to the preindividual, whence there emerges another individuation, and so on. Rather than individuals with essential qualities, we have emergent individuated beings always on the verge of dissolution.

For Stiegler, however—and this is where he moves beyond Simondon—this open quality of individuation is tied to the technical, prosthetic composition of the preindividual milieu—which is to say, in Stiegler's terms, to epiphylogenesis. The "individual's inadequacy to itself," he writes, is caused by the fact that—as per the first volume of *Technics and Time*, but now with reference to "the incompleteness of the individual and not merely of the species"—"the individual always already interiorizes its exterior environment" in its coindividuation with technical forms (2015c:70–71). Linking the metastability of the preindividual milieu to the epiphylogenetic "already there" of technics, Stiegler describes prostheses as consisting in "*originary disequilibrium*" (2009b:55). Combining the coconstitution of human beings and technical forms with the imbrication of psychic and collective individuation, Stiegler thereby models what for him is the inseparability of psychic, collective, *and technical* individuation.

This is Stiegler's criticism of Simondon: that he failed to note the entwinement of technical forms in the processes of psychic and collective individuation, both as forming the preindividual milieu and as themselves coindividuated within these processes. Technics, for Simondon, claims Stiegler, is "merely a moment in the process of psychic and collective individuation, and plays no role in the constitution of preindividual milieus." In Simondon's account, then, "the pre-individual is not techno-logical" (Stiegler 2006:325; my translation). By contrast, Stiegler draws on his understanding of the prosthetic development of humankind to argue that "the technical milieu which is the condition for the encounter of the *I* and the *we*: the individuation of the *I* and the *we* is, in this sense, also the individuation of a *technical system* (something Simondon strangely didn't see)" (2014:51). This is the reason for the resemblance we noted earlier between epiphylogenesis and psychosocial individuation: the technical milieu supports the epiphylogenetic forms that permit the effectiveness of the preindividual milieu. Thus, for Stiegler, the individuation of technical forms both frames and is entwined with psychic and collective individuation, "as the play between this technical system, produced by technical individuation, other social systems, produced by collective individuation, . . . and the psychic system, produced by psychic individuation" (2006:328, my translation)—in other words, "a single process of *psychic, collective and technical* individuation" (2014:51).

Drawing together his readings of Leroi-Gourhan and Simondon, then, we see that Stiegler has an understanding of human beings—individuals, collectives, and humankind—as irreducibly coconstituted with and through their technical forms. As the human for Stiegler is always constituted along with its technical prostheses, it is always in and through its relation to these prostheses that its capacities are developed, in a spiral of reciprocal internalization and externalization in which interior and exterior cease to be metaphysically separated and become regions of a dynamic process.

TOWARD THE DECISION

This spiral brings us to what for our purposes is the key dimension of Stiegler's thought. Crucially, his account of the transductive relation between human beings and their technical forms includes a reflexive moment, in which those psychosocial beings coindividuated with technical forms evaluate their relation to these forms and make decisions about the mode of their adoption while remaining on a common plane of immanence. This is one meaning of the open-endedness of individuation: the emergence of psychic, collective, and technical beings from the technically constituted preindividual milieu is only one moment in the ongoing flux of individuation and disindividuation, crystallization and deliquescence. The reflexive moment in which psychosocial beings evaluate their relation to the technical beings with which they find themselves coindividuated is situated by Stiegler as a moment within this flux; it is a purely immanent decision.

We have already seen that for Stiegler, following Leroi-Gourhan, the human brain develops in tandem with tool use. In this sense, for Stiegler, the reflexivity of consciousness, far from guaranteeing human ontological transcendence, is constitutively technical. To describe this reflexivity, Stiegler develops his own use of the term *epokhè*, or suspension, which he attributes to the stoics, Husserl, and Heidegger (see Stiegler 2003) and elaborates into a process he dubs "epohkhal redoubling" (2009a:60). As its name suggests, this process is composed of two moments; though these might or might not be chronologically distinct, what is most of interest in their relation is its recursive structure.

The process goes as follows. A new technical form suspends, to some extent, the operation of the previous regime: this is a first *epokhè*, a first suspension or interruption, in the process of technical development. This first suspension might be followed or accompanied by a second: the moment at which individuated psychosocial beings turn toward and suspend the effects of this new technical regime with which they have been coindividuated and, in this hiatus, decide which of the tendencies of this regime they will seek to promote. In the understanding of *pharmacology* he derives from Derrrida's reading of Plato's *Phaedrus* (see Derrida 1981:61–171), Stiegler sees any given technical form as neither good nor bad in itself but as offering both harmful and beneficial tendencies. From Derrida's essay, Stiegler takes the key connection between the epokhal crisis occasioned by a new technical regime (in this case, what Socrates describes as the Sophists' abusive substitution of the artificial, external memory housed in the disruptive technology of writing for true, internal memory) and the ability of the technical form in question to operate in this crisis as either poison or cure, its inseparable offering of both a tendency and a countertendency. And the decision on which of these tendencies to promote is what takes place at the second moment of immanent interruption within the process of psychosocial and technical individuation. Stiegler describes the process as follows: "If *tekhnè* suspends the programs in force, then knowledge also returns to suspend all stable effects, *tekhnè*'s 'repercussions,' by redoubling them. This is "epokhal redoubling" (2009a:60).

This "return" on the part of "knowledge" is key to Stiegler's role in developing our overall model. It is his immanent reflexive moment, the "second epokhal redoubling," that can suspend the determined, disruptive effects of a new technical regime (itself a partial suspension of the previous regime) and decide to promote the beneficial adoption of its technical forms. Stiegler gives a thorough account of this process (in which the connection to knowledge is especially clear) in *Automatic Society*: "A technological epokhe is what breaks with constituted automatisms, with automatisms that have been socialized and are capable of producing their own dis-automatization through appropriated knowledge: the suspension of socialized automatisms . . . occurs when new, asocial automatisms are set up. A second moment of shock (the second redoubling) then produces new capacities for dis-automatization" (2016:12).

It is crucial to stress that Stiegler thinks this second moment as recursively immanent to the process of technological change; otherwise, it might all too easily resemble a renewed transcendence on the part of human consciousness of what in such an understanding would become mere instruments. Stiegler addresses this risk directly, in fact, in the second volume of *Technics and Time* (here using his terms "the *who*"—psychosocial beings—and "the *what*"—the "already there" of their epiphylogenetic inheritance):

> The operation of epokhal redoubling (the addition of a new programmatic level partially suspending previous levels' effectiveness)

> is a passive synthesis; it is also the genesis of the *what* in general. The second epokhal redoubling, an "appropriation" by the *who* of the first redoubling, is an "active" synthesis. But we must use these quotation marks when referring to appropriation and activity, insofar as this redoubling of redoubling is always already in the process of clearing a new path for the technical tendency and for a new stage of passive synthetization. (2009a:96; translation modified)

This is a moment of "appropriation," of "active" synthesis, to the extent that it entails evaluation and decision. As the appropriation of a constitutive dimension of heteronymy, it might be compared with an existentialist moment of assumption, in which a subject realizes her freedom by choosing the situation in which she happens to find herself. These terms are misleading, however, to the extent that they imply the framework of subject-object relations, precisely, with its mutual transcendence of interiority and exteriority—as the existentialist version, with its celebration of uniquely human transcendence-through-projection, makes amply clear. Hence Stiegler's insistence on his quotation marks.

The care Stiegler takes here to avoid such misunderstanding shows the importance he attaches to the processual immanence of this reflexive moment and the originary intertwining of "interior" and "exterior." To miss this processual immanence, he argues, would be to miss the articulation of the *who* and the *what*, and so to miss the precise margin for decision available to the *who*. "Technics engages in selection as double redoubling: redoubling of tradition through the technical tendency of which it is the provisional ground, as a diffracting medium, and redoubling of the technical tendency through the *who* that anticipates out of possibilities concealed in this suspension, and inscribes it in memory in details, singularly, transforming it—programmatically"

(2009a:76). "Programmatically," psychosocial beings do enjoy some room for maneuver in relation to their accompanying technical forms—but this possibility is not itself of their making. And, crucially, the converse is also true: there is nothing outside of programming—but there are degrees of determination and indeterminacy. Stiegler writes, "The 'natural' memory of the epiphylogenetic being who is not always already artificial does not exist, having been produced by programs that are largely memory's prostheses. There is *only* that. And the *who*, in its indetermination, programs itself" (2009a:186).

The relative indetermination of the *who* within epokhal assumption is accordingly what allows Stiegler to conceptualize the decision as a moment of reflexivity without transcendence. As he puts it, "I put forth my capacity for individuation . . . insofar as it is inscribed at the heart of a *process* that invents itself and in which I attempt to *participate* as an inventor" (2009b: 48). This is what Stiegler means by describing the philosophical conception of metastability as Simondon's central contribution to the philosophy of freedom (2015b:75). From out of a situation of long-term irresolution, and within an open-ended process of individuation, there can emerge both individuated psychosocial beings and a critical moment of decision. The capacity to decide is not given a priori as the faculty of a metaphysically determined *anthropos*, its exceptional combination of *freedom* and *will*; it emerges in the process of coindividuation, in the rhythm of double epokhal redoubling. Freedom here is not a dream of pure autonomy, of being one's own first cause, as if this were either possible or even conceivable among the finitude and contingency of the sublunary world. It is the possibility, for psychosocial beings, of deciding pharmacologically between the tendencies of the technical forms with which they have been coindividuated. And the will is no longer the categorically given faculty of a transcendent subject but, rather, "what is secreted by a *process* of psycho-social individuation" (Stiegler 2011:96; translation modified). The decision, in which freedom and will come together, thus appears less as a sublimely sovereign verdict from on high than as an orientation within a process. In keeping with the results of some recent neuroscientific research into decision-making, for example (see Hayles 2017:7), we can think of the Stieglerian decision as a matter of promoting a tendency within the flow of its own emergence and evanescence.

As Stiegler writes, again foregrounding the articulation of determination and indeterminacy, "Technics . . . constitutes my indeterminacy, that is to say my inadequacy to myself, precisely as determination. And the point is that everything related to a problem of decision proceeds through a structure of this kind, as dynamic coupling" (2003:163). If the decision is always related to exteriority, this is the *relative* exteriority of the technical forms to be adopted. Accordingly, "It is *always* in the play of exteriority that decisions are made. Questions which call for decisions arise only like this: they always go through exteriority, they are induced and led, more or less palpably, by exteriorisation" (163). These decisions are made, then, in the moment of second epokhal redoubling, that reflexive moment when psychosocial beings turn to consider their relation to the new technical forms that have, to an extent, suspended a previous technical regime. Hence, for Stiegler, "there is decision, strictly speaking, when there is *epokhè*" (160); the decision operates as a suspension but not as transcendence (as it does in the decision of the Schmittian sovereign to suspend the rule of law, for example). Stiegler uses *epokhè* to figure the decision as a processually immanent suspension, the exercise of relative freedom in and as a relative hiatus. The fantasy exteriority of the exceptional sovereign is replaced by the relative exteriority of individuated being and milieu; the fantasy transcendence of human-exceptionalist reflexive consciousness is replaced by the reflexivity without transcendence of the pharmacological decision. For Stiegler, then, *epokhè* means this: heteronomously constituted along with the technical forms with which they have emerged in a process of coindividuation, psychosocial beings turn immanently back on these forms from within this process, suspending to an extent the determinative authority of the new technical regime in a decision to promote this or that of its tendencies.

In relation to the decision, Stiegler's approach to this relative exteriority, and the recursive loop of assumption in which its two terms meet, is most strongly informed by a figure he finds in the work of Gilles Deleuze: the *quasi-cause*. Behind Deleuze, we of course find Nietzsche: specifically, the Nietzsche of *amor fati* and the eternal return, also, of course, a definitive influence on that existentialist model of assumption. In Zarathustra's magnificent style, the recursive loop we are following here is expressed as follows: "All 'it was' is a fragment, a riddle, a dreadful accident—until the creative will says to it: 'But I willed it thus!' " (Nietzsche 1987:163; translation modified) Neither Stiegler nor, indeed, Deleuze will inflate the act of choice quite so gloriously in relation to the hinterland of its constitution; for them both, however, Nietzsche's conjunction of the accident and the decision will remain crucial. Indeed, it is in his adoption of Deleuze's quasi-cause that Stiegler's importance as a thinker of the accident is located, precisely inasmuch as he thinks this *in conjuction with* the decision.

The quasi-cause is in a sense Deleuze's way of theorizing a situated choice—indeed, an act of will—without importing the existentialist mechanics of ontological transcendence that are one of the later avatars of Nietzsche's magnificent, first-person "creative will." Hence, if the Deleuze in question is certainly operating in this broadly Nietzschean context, his primary reference is the Stoics—which is to say, we might recall, one of the averred sources of Stiegler's *epokhè*. If stoical *epokhè* is for Stiegler "decision as the interruption of a flow, an arrest or halt that issues a judgment" (2003:160), what matters most to Deleuze is the relation of this gesture to *sense*, its suspensive capacity to extract from an event a new dimension of significance. Deleuze divides the event in two: on one side, the contingent occurrence, what happens to have happened (which Deleuze, like Nietzsche, calls the *accident*); on the other, the impersonal, preindividual, incorporeal, "neutral" sense expressed by this occurrence. The task of the individual to whom the event has occurred is to double the actual occurrence with a "counter-actualization" capable of extracting this sense; as Deleuze puts it, of "attaining this will that the event creates in us; of becoming the quasi-cause of what is produced within us" (1990:148). Rather than lamenting the accident as the incursion of a malign external force (this way lies only the impasse of *ressentiment*), I inhabit it even more deeply, moving inward to find the sense it expresses, which overflows this contingent occurrence and with which I can access energies and forces within, beyond, and, if need be, against this occurrence. And so, says Deleuze, "war is waged against war" (149).

This is where the quasi-cause meets Stiegler's pharmacological model: the attempt to counter-actualize the accident takes the form of the struggle to promote the curative side of a newly disruptive technical form. As Stiegler says of the quasi-cause,

> It is a position with respect to life that consists in positing that what wounds me, what weakens me—if it does not kill me—is also my chance. My chance lies only there; it is not providence—transcendental or transcendent—that will save us. There is no

> salvation, and it is not a question of being saved, but of being worthy. This is what is magnificent in Deleuze.
>
> In fact, it is always a question of being worthy of technics—that is, of the accidental—including when it leaves me destitute. To be worthy of the strength that it gives me is pretty easy. But to be worthy of what it takes away from me is a true test. (2018b:142)

Stiegler is here alluding to the lines with which Deleuze accompanies the definition of the quasi-cause: "Either ethics makes no sense at all, or this is what it means, and it has nothing else to say: not to be unworthy of what happens to us" (1990:149; translation modified). Translated into the idiom of Stieglerian pharmacology, this becomes the attempt to become worthy of the technical form with which we happen to have been coindividuated by promoting its adoption in curative rather than toxic mode.

We can see, then, that Stiegler's use of Deleuze's quasi-cause offers another way to formulate the process of double epokhal redoubling: the shock and suspension of a new technical regime correspond to the contingent occurrence of the accident, and the pharmacological decision as to the mode of adoption of new technical forms corresponds to the retrospective transformation of the accident into a necessity that can be embraced. For Deleuze, the quasi-causal transformation of the accident effects, above all, a change in its *sense*; in Stiegler's terms, this underlines how the decision as to the mode of their adoption marks a commitment to a *value*: the value of the technical forms in question but especially the value of the form of life this mode favors. Hence Stiegler's emphasis on becoming *worthy* of the accident; it is a question of identifying and promoting what in the accident offers positive value and of committing to the future of this value by seeking to promote it.

This evaluative dimension is signaled in part by Stiegler's allusions, around his use of the quasi-cause, to the Nietzsche who, as we saw, stands upstream from Deleuze's formulation and whose transformative "creative will" intervenes to overturn an existing regime of values. When Stiegler writes, for example, that the "blind, fateful necessity" of the accident is quasi-causally inverted (2016:70; translation modified), this dynamic is part of the pharmacological struggle to promote "what makes life *worth* living" (as in the title of 2013b; my emphasis): to affirm within the flux of coindividuation the value of a mode of adoption that could break with a current, dominant regime and generate possibilities the collective in question holds to be beneficial.

With Stiegler's account of double epokhal redoubling as decisive, quasi-causal transvaluation of the accidental emergence of a regime of technics, then, we encounter a discontinuous change that remains immanent to an overall system or process. This change is never simply the doing of the individuated psychosocial beings in question. The relative indetermination of these beings allows them to make decisions to promote this or that tendency, but it is within the process that the discontinuous changes in question occur. The action in question, however, should still be considered sharply interventionist and normatively partisan. Through its pharmacological promotion of a given tendency, the decision on adoption seeks to make a drastic difference to the life elaborated alongside its accompanying technical forms and, in so doing, affirms the value of the tendency it promotes. As Vesco puts it, recalling the importance here of Simondonian metastability, at stake is "precisely the notion of a pause that is both an interruption and the precipice of reconfiguration" (2015:101). And just as metastability opens the possibility of immanent strong discontinuity, so does Stiegler think transvaluation.

The great contribution of Stiegler's model here to our efforts to conceptualize a form of decisive distributed political agency is thus that he manages to think the decision as at once processually immanent and sharply interventionist—as both inflection and interruption. Moreover, the pharmacological choice of value at work in this decision gives it a strongly futural orientation: asserting belief in this value as beneficial, it is also a normative commitment and a partisan prescription.

THE PHARMACOLOGICAL POLITICS OF ADOPTION

So far, then, we have followed those aspects of Stiegler's thinking that allow us to posit decisive political action without the need for any appeal to mythical human exceptionalism. What we have found in Stiegler so far is the *structural possibility* of the decision as not only processually immanent but also retroactively productive of drastic change (in the meaning of its situation) and futurally committed (in its assertion of a value). This structural possibility already allows us to give an account of distributed political agency as decisive in the full sense of the word. In addition to this structural sense, however, Stiegler also thematizes the decision as itself political: indeed, the transformative decision on adoption is the very heart of Stiegler's understanding of politics.

If this is the case, it is because the decision on adoption is always a struggle, to the precise extent that it is also always pharmacological; the decision is a choice between competing values, a quasi-causal struggle to define what counts as beneficial and what counts as harmful. The question of adoption is a matter of strife, of public dispute and struggle: "It is in this way that technics *constitutes* the *problem* of *decision*" (Stiegler 2003:159). The transformative decision on adoption that responds to the disruption introduced by a new technical regime is taken against a horizon of strife; it is always, for Stiegler, a *polemical* decision.

In order not to give the misleading impression that his account of these processes is wholly abstract, and to see what the futural commitment of the decision on adoption as a struggle about values might look like in a particular situation, I will now spend a little time considering some of Stiegler's more polemical writing. Specifically, I will explore his discussions of what he terms "the industrialization of memory" and the widespread automation of work. I will conclude this section with reference to his discussions of the planetary crisis indicated by the notion of the Anthropocene.

Given Stiegler's original concentration on the question of time, and especially the epiphylogenetic transmission of the impersonal memory embedded in technical objects, we will, then, find our first example of the political struggle over the mode of adoption of technical forms in his pharmacology of memory and consciousness. In the second volume of *Technics and Time*, Stiegler pursues this question with reference to what he calls the "industrialization of memory." Recalling his own use of Derrida's reading of Plato on writing, he reminds us that to remedy its finite capacity, human memory is supplemented by all manner of technological supports—and as we learned in *Technics and Time, 1*, this is originary. "Technics does not aid memory," writes Stiegler; "it *is* memory, originarily *assisted* 'retentional finitude' " (2009a:65). This originary technicity exposes memory to the possibility of industrial appropriation and exploitation, when the supports that are "the very conditions of its *e-laboration*" (8) are integrated into an economy of standardized industrial production for commercial gain. "Prosthetization impacts what

Kant in *The Critique of Pure Reason* calls the syntheses of apprehension, reproduction, and recognition. But this impact is possible today only because it is originary" (98). Stiegler accordingly defines the industrialization of memory as "*the industrial synthesis of retentional finitude* subjected, as pre-judgment, to the specific criteriology of calculable credit as the operator of economic development" (9). Writing in the 1990s, he describes memory itself as having become "the primary matter of industrial activity" (61). Two decades later, it would be fair to say that our defining technical regime has hardly given the lie to Stiegler's analysis.

The political dimension of this process lies in this question of criteria, in that this itself encodes a decision on adoption. For Stiegler, "memory is always the object of a politics, of a criteriology by which it selects the events to be retained" (2009a:9). With organic retentional finitude supplemented by programs whose criteria of selection are defined in terms of monetization, memory is constituted according to financial priorities. In the struggle over the value of the technical forms in question, the decision to adopt these according to the criterion of calculable financial profit has recently proved dominant. There thus emerges today what Stiegler describes as "a politics of memory encountering the resistance of an economic imperative" (100). Stiegler's view is that this encounter constitutes a technological event of unsurpassed significance. Glossing Leroi-Gourhan, he writes, "When the elements of a contemporary technics designed for information processing develop, a truly automatic activation of memory will have appeared, as the harbinger of an exteriorization process of the cerebral cortex's functions and, on a wider scale, of the entire nervous system" (79). (Stiegler is moving quickly here: we should not forget that "exteriorization" is a relative term, one pole in an ongoing dynamic spiral of coconstitutive terms.) If it is always the case that, with reference to memory (which by definition is constituted in relation to inherited technical forms), "the question . . . regards effective means of access, and, through this, of sequencing," this question "is being posed very intensely today because the exteriorization of the human as such has reached its limit, meaning that contemporary technical *epokhality* is radical, equal to the most powerful ruptures that humanity has ever known" (75; translation modified). The rupture provoked by the digital "exteriorization"—i.e., coconstitution—of human capacities would appear to be not only unsurpassed but unsurpassable, if this "exteriorization" really has here reached its limit.

Although Stiegler might appear in such moments to be aligning himself with a techno-catastrophism, his understanding of technics as originary prostheticity in fact rules out any such apocalyptic bent. For what responds to epokhal rupture is always, one way or another, a political decision concerning the way in which the newly coconstitutive technology will be inhabited. Faced with "a truly automatic activation of memory," we are nevertheless not looking at the usual fantasy of the "rise of the machines." If "this is clearly a question of the autonomization of *tekhnè* and its automobility," we nonetheless know "that there has always already been 'autonomization' of the techno-logic automaton," and therefore that "it is not a question of an autonomization relative to a golden age of the mastery of a technics 'closer' to human beings, but rather of a becoming-hegemonic of developmental economic imperatives in accord with a particular interpretation of time qua value" (2009a:140). This is not destiny; this is a particular mode of adoption, which might accordingly be challenged. Or not—and this is precisely the pharmacological struggle over value, in all its polemical force.

"Then the question remains of who regulates selection" (2009a:128). In the face of the alliance between speculative finance capital and the psychotechnologies working to confine memory and consciousness to programmable consumerist habits (2010b:128), or what he calls this "de facto economic psychopower," Stiegler is looking for a "psychopolitics" that might quasi-causally transform this economy into "a true ecology" (35). Such an ecology would entail "an industrial but ecological economy of cognitive and affective functions forming a new civilizational model on the basis of a reorientation of our contemporary industrial reality" (2013a:126). Stiegler repeatedly stresses that the attempt to develop such an alternative economy must be understood as a struggle. Indeed, he often goes further, phrasing this struggle in the language of war, in uncompromising terms. As a form of "total war," in *Symbolic Misery, 1* (2014:10), or in *What Makes Life Worth Living*, by

> an economic war without mercy: a concealed conflict, a bottomless hypocrisy, a constant struggle, exhausting the Earth and its inhabitants, and leaving a billion of them in abominable economic misery while ruining the whole of the human world more quickly and ever more irreversibly, such that, in this war disguised as peace, it will not be long before everyone loses. The name of this war is globalization—a globalization in which industrial technologies have become weapons that destroy ecosystems, social structures and psychic apparatuses. (2013b: 9)

If this "total war" is thus at once economic, ecological, and technological, Stiegler's conception of any possible strategy starts with understanding both the specific technical forms that define our time (say, from the financial exploitation of consciousness to the biotechnological use of molecules for data storage) and the adoption of this technicity in the interests of speculative finance capital. Such understanding thus forms part of the moment of second epokhal redoubling: from within the disruptive shock of this technical regime, Stiegler's pharmacology calls us to reengage this constitutive technicity by inventing new, curative economic and industrial politics and policies that would, by contrast, promote the beneficial adoption of these technologies.

Here we see the quasi-causal dimension of this pharmacology. Rather than bemoaning the inherent evil of new technologies, Stiegler wants us to look more closely at these technologies and identify within them the necessary curative power. (Along with Derrida's Plato, Heidegger's Hölderlin—"But where the danger is, grows | The saving power also" [Heidegger 1993:340]—is certainly on his mind at such moments.) In this case, Stiegler finds such a positive pharmacology of counter-effectuation in those areas—notably, digital activist communities, or alternative forums for education and argument—already adopting digital technologies in ways that promote their beneficial tendencies (2015b:40). The challenge, in turn, would be to generalize this counter-effectuation; only with such generalization do we find the collective dimension essential to the moment of second epokhal redoubling.

That this second stage of epokhal redoubling requires a specifically collective decision comes through especially clearly in Stiegler's account of our next substantive question; namely, the problems likely to result from the massive and widespread automation of paid work. In the first volume of *Automatic Society*, Stiegler starts his analysis by setting out the nature of the changes in question. Citing reports of various pieces of research, he notes the potentially devastating effect of large-scale computerized automation on the landscape of employment in developed industrial societies, with approximately 50 percent of jobs highlighted as potentially at risk over the next two decades. In one article referenced by Stiegler, for example, Carl Benedikt Frey and Michael A. Osborne write, "According to our estimate,

47 percent of total US employment is in the high risk category, meaning that associated occupations are potentially automatable over some unspecified number of years, perhaps a decade or two" (2013:38). In the developing world, the scenario is even bleaker, with the projected rate of loss ranging from 55 percent to 85 percent (Citi GPS 2016). To clarify what follows, it should be noted that Stiegler's focus remains on developed industrial societies.

Responses to this scenario typically maintain a metaphysical opposition between human and machine in either utopian or apocalyptic mode: a vision of vast hours of newly rewarding leisure time (as evoked especially by Marx in the "Fragment on Machines" from the *Grundrisse* [Marx 1973] and more recently promoted by Srnicek and Williams [2015]), or, alternatively, of "the machines taking over," with the machine as, respectively, the still-loyal servant or the rebel servant become master. As we have seen, even when apparently aligned with a techno-catastrophism, Stiegler's spiral of coconstitutive internalization and externalization is designed to rule out this metaphysical opposition and the relation of instrumental exteriority it imposes. For all its fantasized transcendence, such exteriority can only, in his view, prove disastrous. This is accordingly the framework within which he explores the question of automation.

The heart of the pharmacological is, as Stiegler puts it, the fact that "that which constitutes—technics—destitutes. Technics is at once and in the same gesture that which constitutes possibility and that which bars possibility" (2018b:141). When a technical form is adopted in ways that bar possibility, the resulting scenario is what Stiegler terms *proletarianization* (borrowing the term from Simondon). In this version (derived, indeed, from Marx's "Fragment on Machines"), proletarianization is defined not by having nothing to sell except one's labor power but by a loss of knowledge—or, rather, it is this loss that transforms labor power from a form of knowledge into a commodity. As a result, the laborer becomes the servant of the machine, in which the knowledge in question is now functionally located (see Stiegler 2010a:35–39). So in Stiegler's pharmacological model, proletarianization is the harm that results when a technical form is adopted in such a way as to exclude those who use it from participation in the production and circulation of knowledge on the basis of a functional understanding of the technical forms defining the current regime.

This scenario is of particular importance to Stiegler because in his view, it describes the situation of "generalized proletarianization" (2010a:35), which defines our current technical regime: "what Socrates describes in *Phaedrus*, namely that the *exteriorization of memory is a loss of memory and knowledge*, has today become the stuff of everyday experience in *all* aspects of our existence" (29). In *Automatic Society*, as elsewhere (particularly *Symbolic Misery*), Stiegler expresses this through a vocabulary derived from Husserl, which distinguishes between primary retentions (immediate memory of what passes through consciousness), secondary retentions (memory of a previous experience), and—this is Stiegler's addition—tertiary retentions (epiphylogenetic technical supports through which memory is inherited). The "employee," for Stiegler, the wage laborer, is by definition proletarianized, alienated from the individual and collective knowledge formed by primary and secondary retentions. "Employment," he writes, "is characterized by the fact that the retentions produced by work no longer pass via the brains of the producers, who are themselves no longer individuated by work, and who are therefore no longer the bearers or producers of work-knowledge" (2016:160–61).

The employee's sole function in this situation is to optimize the performance of machines that "utilize collective secondary retentions . . . turned into machinic tertiary retentions or technologies of all kinds" (2016:162). The situation is exacerbated by the cybernetic management of labor processes via algorithmic feedback. If the employee is initially "subject to protocols supported and defined by machines, . . . reporting systems and management control," she is now literally "without employment, short-circuited by algorithms that outstrip psychic individuals and collective individuals" (161). What is more, the increasing automation of labor, combined with this algorithmic harvesting of data, sees the unemployed employee mutate into a producer of financially translatable information. Stiegler describes this via the concept of "algorithmic governmentality," which he takes from Antoinette Rouvroy and Thomas Berns (2013): "Increasingly, systems of production and distribution characteristic of algorithmic governmentality will be regulated according to trace-producing *externalities* generated by networked individuals, these systems being ever more closely integrated, ever more autonomous, and served no longer by producers but by consumers—to the extent that any of these are left amidst the generalized bankruptcy this system inherently installs" (Stiegler 2016:163; translation modified).

From worker to machine-operating employee to data farm animal, the trajectory of proletarianization Stiegler traces has its milestones in key moments of lost know-how, as the loop of internalization and externalization is severed and the knowledge embedded in technical prostheses is not functionally "appropriated." At no point in this process are we dealing with an activity meriting the name of *work*: for Stiegler, this term is valid only where such "appropriation" is operative. By contrast, what we see in this process is only *employment*, up to and including the harvesting of data generated by unemployed employees. Its first two phases comprise full industrialization with "scientific management" à la Taylor, followed by the full extension of the consumerist economy. Through these, "proletarianization transforms work *in its totality* into jobs emptied of all knowledge and requiring only skills that represent 'employability,' that is, 'adaptability' " (2016:172; translation modified). Its third phase, through which we are currently living, sees the activities of consumers start to displace labor in a conventional sense as the source of profit. If this is often termed unpaid "work" on the part of these consumers, Stiegler insists that the term has no place here. Far from work, this is the "*employment of unpaid time*—that, by *harnessing this employment*, . . . feeds, reinforces and sets the parameters of the collective retentions automatically and performatively produced by this *totally* computational capitalism" (163; translation modified; on this, see also Fuchs 2016). Already in the era of analogue broadcasting, "Liberated time that does not become free work becomes available brain time for the fabrication of mass consumer markets" (179; translation modified; and for the origins of the phrase "available brain time," see Stiegler 2015a:46). The shift to digital platforms has only found more capillary ways to monetize this available brain time, now *employed* to generate the data that drive ever more sharply targeted forms of commercial address. Indeed, so capillary are these ways that they extend beyond the brain, harvesting in new form the output of the human-technical composite that we have always been.

In the face of such a scenario, it is important to emphasize again that Stiegler is not describing an ever-worsening process of "alienation" as a result of the evils of mechanization and automation. The loop of internalization and externalization renders such a concept of alienation inoperative here. As we have seen, the question can be only one of *how* this relation is to be adopted—and most particularly, how the knowledge it produces is to be appropriated. Consequently, Stiegler's position in relation to the progressive replacement of the worker by the unemployed data farm animal is not one of exclusive despair; pharmacologically, there is hope here, too, inasmuch as the values driving this process are still to be fought for. Whereas the metaphysical opposition between human and machine can see increasing automation only in terms of the reduction of labor time, Stiegler insists that the stakes are in fact "*the end of employment* and . . . *the reinvention of work*" (Stiegler 2016:166). The question, he writes, is whether "increased productivity, which will escalate in spectacular fashion in the next few years thanks to full and generalized automation, should *liberate time* or *liberate work*" (170; translation modified).

If the former possibility simply makes even more brain time available for commercial exploitation, the latter offers the possibility of a very different, *contributive* economy, through which work in the strong sense—entailing participation in the cycles of reinternalization of exteriorized knowledge—might once again become possible. This is the political decision as to the meaning of automation: given that purchasing power is likely to collapse with the massive replacement of wage labor by automation (2016:163), through what economic forms are the resulting productivity gains to be redistributed? Assuming that this redistribution itself is held to be desirable, of course—the alternative plainly being the mass destitution of former unemployed data farm animals whose data have become worthless in the absence of their purchasing power. The strife between these alternatives is the form taken in our time by the struggle to define the second moment of epokhal redoubling.

For those who want a future that would not feature such mass destitution, the challenge would be to frame a "right to work" in which work is defined "as it develops outside of employment, and as the *power of dis-automatization*" (2016:166' translation modified). This will not be done by leaving everything to the third sector or by the creation of a universal basic income (however important this latter step might be). Stiegler's specific proposal here is for an economy founded on what he calls a "contributory income" (180), in which the free time produced by massive automation would allow groups and individuals to develop forms of technically constituted knowledge the social contributions of which would be remunerated. The economics of this model rests on the notion of "intermittent contribution" (according to which French artists have been able to receive state funding even when not actively working); its psychosocial potential lies in its integration of those involved into the processes that, for Stiegler, define *work* in the strong sense—that is, the production and circulation of material and symbolic goods on the basis of functional technical knowledge. As he writes, accordingly, "the politics of work reinvented in this way [constitutes] the cornerstone of a *general politics of recapacitation*" (181; translation modified).

Stiegler's call here for immediate local experimentation in this sense, in which "contributory research" would be allied to specially developed "contributory technologies" (2016:181), came in the context of his work to establish in the Seine-Saint-Denis region a research initiative that would run on exactly these lines. The initial activities of this Territoire Apprenant Contributif, based in the *Établissement public territorial* of Plaine Commune, were then described in *The Neganthropocene* (see Stiegler 2018a:123–37). Allowing for Stiegler's adoption at this time of the language of "exosomatization," entropy, and negentropy (derived variously from Alfred Lotka, Erwin Schrödinger, and Nicholas Georgescu-Roegen), we can clearly see how he framed the action undertaken at Plaine Commune as a form of quasi-causal counter-actualization in the service of second epokhal redoubling. Taking the "reticulated" structure of digital networks as a key dimension of the technical form whose disruptive effects currently give rise to the need for its curative adoption, this action looks to develop this same reticulation in a way that is productive not of monetizable information but of individuating knowledge:

> Our ambition at Plaine Commune is to become a model in the field of "truly smart cities," by reticulating this territory on the basis of completely new data architectures that are profoundly socialized at every exorganic scale. . . . This process of reticulating this whole territory . . . aims to enable *deliberative* processes to be set up at *all* levels, and to be articulated with one another—because they will share the same deliberation protocols. Our intention is to use these deliberative processes to generate local forms of knowledge capable of engendering negentropy—to place automatisms at the service of dis-automatization. (2018a:127)

What is more, Stiegler emphasizes that the local gains of such repurposing will have to be scalable, as required "by the *shift from the regional micro-economy to the new macro-economy*" (137). Given that the regime to be confronted is operative on a planetary scale, it will be necessary to develop alternative modes of adoption to combat its toxic effects on this same level. If this is evidently ambitious, my aim here is not to evaluate this ambition; the point for our purposes, rather, is that, together with the arguments of *Automatic Society*, the work undertaken at Plaine Commune offers the most detailed example we have of how Stiegler envisioned the struggle characterizing the moment of second epokhal redoubling as the fight for a beneficial mode of adoption.

In moving to *The Neganthropocene* to discuss Stiegler's work at Plaine Commune, we have necessarily begun to feature some of the language of negentropy that dominated his later idiom and, notably, his discussions of the Anthropocene. To conclude this section on Stiegler's pharmacological politics of adoption, I propose now to consider something of what was at stake for Stiegler in this language, in particular its relation to the question of value as this motivates the quasi-causal struggle for a second epokhal redoubling.

Stiegler's neologism, the *Neganthropocene*, punningly combines negentropy and Anthropocene to signal his claim that the crisis identified in the notion of the Anthropocene is a crisis of value. ("Anthr-" and "entr-" are indistinguishable in spoken French, so Stiegler's pun is rather less forced in its original language.) Broadly speaking, in this thinking, he imports the conventional understanding of the Anthropocene as an era defined by human activity having become the dominant geophysical force on a planetary scale (2016:216–17). More particularly, he characterizes this human geophysical dominance in terms of "the development of the industrial way of life [which] has become toxic not only on the plane of minds and libido, but also on the geophysical and biological plane" (2013b:88). Drawing together his analyses of the industrialization of memory and the automation of work, the Anthropocene is for Stiegler the time of a form of hyperindustrial capitalism "founded on the industrial exploitation of modelled and digitalized traces" (2016:15), in which calculation is the dominant mode of rationality (most recently in algorithmically automated form) and projection is confined to the anticipation of already calculated differentials. This is the connection to entropy: if such a regime of digital-algorithmic calculation necessarily works by differentiation (and might hence be thought to be negentropic), the differences in question are defined by their assimilation in advance into processes of calculation, as "information—that is, data, which can present itself only as formatted in terms of its a priori calculability" (2017:81). What this eliminates is accordingly the singular, the incalculable, or what Stiegler elsewhere, with Heraclitus, calls the *anelpiston*, the unexpected and unhoped-for (2015a:164). This is why the question posed by "the entropic catastrophe that is the Anthropocene" (2016:15) is the question of value. As Ian James writes, "The Anthropocene, for Stiegler, does not just describe the embedding of lasting anthropogenic signals within the geological record but rather concerns the embedded value system of our technoscientific age" (2019:209). Understanding it in this way as the concrete form of general equivalence, Stiegler describes the Anthropocene as the planetary strife between entropy and negentropy (2016:33), in which the former is so far emerging as dominant. As indeed it would have to, if this were simply thermodynamics.

But we are not solely in the domain of thermodynamics, of course—and this is precisely the question: we are in the realm of politics, which is to say, of conflict over values. In Nietzschean style, our quasi-causal tasks are therefore to think the Anthropocene as a geologic era realizing the nihilistic devaluation of all values and, accordingly, to devote ourselves to the transvaluation of its pseudo-values (2018a:38). This is what Stiegler punningly calls "neganthropy," an ensemble of practices promoting "the value of value" (2016:15; 2018a:46). "Only in this way," he writes, "can and must the passage from the Anthropocene to the Neganthropocene be accomplished" (2016:15).

As Stiegler wrote in *What Makes Life Worth Living*, "*Krisis* means 'decision.' We all now know that it is the future of terrestrial life that is at stake with unprecedented urgency" (2013b:5). In the fight for the beneficial transvaluation of an existing regime of pseudo-values, then, he insists that what is needed is decisive political action. Indeed, this crisis represents, in his view, "an astonishing renovation of the political question as such" (2018a:78), which demands "a *global geopolitical alternative to transhumanist marketing*" (79). The need for planetary scalability, as seen earlier, means that transvaluation here requires "major economic and political change" (2016:238), "a *new political regime*," and a "macro-economic revolution . . . to make this new regime of individuation solvent" (2018a:82). As Daniel Ross writes, it "demands a reconsideration of the broadest macroeconomic questions and their relationship to the speed and power made possible by the digitalized, networked and algorithmic technical system" (2018:23). The change called for in the name of such transvaluation is thus both large-scale and drastic. For Stiegler, it entails the active production of a decisive rupture; indeed, as part of its refusal of incalculable singularity, the entropic tendency against which it is necessary to fight is defined for Stiegler by its active suppression of any curative break, any epokhal moment (2016:118–19).

Just as he describes individuation as a process taking place via intermittent leaps (2016:183), so here Stiegler maintains the processual and the discontinuous together in a model of intermittent transition (84) that becomes the very focus of his neganthropology. Breakthroughs, he says, are possible "only on the condition of allowing such improbable occurrences to occur, letting them emerge from out of these non-programmable forms of intermittence—even if the most favourable conditions for such an emergence . . . can and must be maintained and cultivated" (224; translation modified). Within continuity, the possibility of discontinuity must be cared for as precisely that which cannot be produced; as this care takes the form of the beneficial adoption of epokhal technical change, the disruptive break made by such change is thereby quasi-causally brought to sustain the possibility of another, creative departure from within the very regime of carelessness inaugurated by this disruption. The fragility of forms means that as the always temporary stabilization of preindividual possibilities, any given form is potentially exposed to the destructive effects of an accelerated mode of disindividuation. In the absence of a quasi-causal reversal, in such scenarios, the accident of coindividuation occurs as disruption. If disindividuation is inevitable, however, disruption is not; from within this same process, it, too, can be interrupted.

The decisive political action for which Stiegler calls is thus a kind of processual discontinuity. As Vesco writes, "Stiegler theorizes situations of decision not as a break with 'what is' but instead as ever-emerging reconfigurations" (2015:101). As we have seen, the figure of metastability as the medium of immanent strong discontinuity—drastic change within a system—allows Stiegler to think the decision on adoption as at once inflection and interruption. As a watershed, say: the inflection of a flow (i.e., of the ongoing process of individuation and disindividuation) whose transvaluative promotion of one of the flow's tendencies effects a suspensive, interruptive, dramatic shift in its milieu. An epochal shift, we might say (using, for once, the resources of the conventional spelling). The absence of any appeal here to exteriority means there is no incompatibility between inflection and interruption: the critical decision comes as a fold with the force of a cut. An "agential cut," in Barad's phrase (1997:140), reconfiguring the elements of its situation as participants *in the act*, committing to the value of *this* alignment.

THE ACCIDENTAL DECISION

The suspensive dynamic of *epokhè* thus allows Stiegler to model his pharmacological politics of adoption as operating in the mode of immanent decisive intervention, in both senses: as proceeding from a decision and as making a major difference within a conflict over value. This is his great significance for our purposes: with this model, he is able to think such decisive intervention as both antagonistically engaged and the locus of futural commitment, while maintaining it as fully immanent to the process of its emergence.

Of course, the forces pulling against this immanence are considerable. This is exactly why Stiegler is so valuable here; the precision of his attention to the problem of a processually immanent decisive intervention is outstanding. Inevitably, though, tensions can become apparent between the respective commitments to immanence and to decisiveness. To an extent, this replays with regard to Stiegler the familiar problem of articulating facts into values, or description into prescription. As Claire Colebrook has suggested, for example (2017), Stiegler's own philosophical commitments can make it difficult for him to ground his normative positions (i.e., that this or that mode of adoption is toxic or beneficial). The question of normativity is, of course, a significant aspect of the model I am developing here, inasmuch as I am describing distributed political agency as crystallizing the existential stakes of its milieu and—thanks to Stiegler—as making a partisan futural commitment within a conflict over value. Despite this, however, such tensions as might appear in Stiegler's writings at moments of normative assertion are not of primary importance for our concerns. It is true that for some critics, Stiegler's normative commitments lead him to reintroduce a dimension of ontological transcendence effectively programmed by his phenomenological inheritance. Most frequently, this criticism has surfaced in arguments around Stiegler's alleged determination of epiphylogenesis as the singular opening of a specifically human form of being. This determination would, it is claimed, see him repeating a phenomenological form of philosophical anthropology. (On this, see, for example, Bennington 1996; Beardsworth 1998; Roberts 2005; Ekman 2007; Colony 2011; and Turner 2016.) Less commonly, Stiegler's attachment to a strong definition of work as the production of value is read as reinstalling a Marxian affirmation of human species-being, with work valorized to the extent that it guarantees the mediation between individual and species by posing the recognition in question as the affirmation of the productive activity and negation of the given world proper to the human form of being. (See, for example, De Beistegui 2013; and for the Marxian version of this affirmation, Marx 1959.)

However, just as Stiegler insists that his interest in epiphylogenesis is indifferent to its supposed speciation (see, for example, 2009a:255n44; as cited earlier), so do I want to insist that these alleged ghosts are of little significance here. They could be of great significance, it is true—if they implied that the quasi-causal decision on adoption could be taken only by ontologically transcendent human beings. But nothing in Stiegler's accounts of this decision requires this interpretation. Indeed, the terms by which he prefers to designate the agents that emerge in such decisions—"the *who*," "psychosocial beings"—serve precisely to distance these accounts from such a reading. Reflexive decisions on adoption certainly are taken by human beings. But Stiegler's understanding of such beings as coindividuated with technical forms means that the form of life we call human is itself already a composite: in all senses of the phrase, these decisions are not taken by human beings *alone*.

What Stiegler has given this study, then, derives from his conception of the quasi-causal decision to adopt the accident as necessity. It is Stiegler who has first allowed us to think the accident and the decision together, configuring this relation in a mode of immanent recursivity that frees the decision from any dependence on the ontological transcendence supposedly secured by uniquely human intentional consciousness, to recast it as reflexivity without transcendence. And so our model of politics, in terms of the decisive, distributed agency of ontologically diverse alliances, gains a key component: the decision in the strong sense, now understood as processually emergent, immanent to its milieu and independent of a priori human capacities. Immanent to the accidental emergence of its agential site, the decision as thought by Stiegler is, moreover, a polemical, partisan intervention within a conflict over value. Its inflection of the accident in a futural commitment to one of its tendencies makes a drastic difference to a state of affairs. Immanent inflection is also political interruption. Thanks to Stiegler, we can thus understand politically decisive action, in the full sense of the term, independently of any human exceptionalism. The next task will be to combine this understanding with the model of distributed political agency established in horizon 1. Horizon 2 is dedicated to this task.

Horizon 2

AT THE SPEED OF THE DIGITAL ALGORITHM

In chapter 2, we encountered one aspect of what we might call digital politics: the consequences for political economy of the massive widespread automation of labor, as enabled by the digital-algorithmic management of the production, delivery, and circulation of goods and services. In this section, we will consider a second: the significance for our core question of political agency of the relatively recent appearance of a digital political culture. By this, I mean not only the partial shift online of institutions of government and civil society but also, and especially, the shaping of political and electoral practices by processes that depend on the capacities of digital algorithms, server farms, and vast processing power. At the planetary scale, the emergence of these processes has taken place against the backdrop of what Benjamin Bratton describes as the "direct conflict" between two forms of sovereignty: on the one hand, "the Modern sovereignty of the State (which would produce one form of public)," and on the other, the "network sovereignty" of ubiquitous computing "that would produce another" (2010: para. 13).

At the level of the polity, the processes in question are duly producing a mutation in the effective constitution of the public. Whether we think of this as identifying an aggregate of autonomous individuals rationally pursuing more or less enlightened self-interest, as a rough amalgam of more or less antagonistic interest groups, or, indeed, as a multitude of bodies subject to statistical management, its—for now partial—recomposition in terms of user profiles has posed fundamental questions to post-Enlightenment understandings of political subjectification. Massive data harvesting and granular voter profiling have redefined the production of citizens through political interpellation. In alliance with more or less legal financial instruments, semi-clandestine government surveillance and manipulation, and the digital weaponization of information, they have given rise to a political landscape whose defining operations of assembly, deliberation, and decision-making take place partly in realms inaccessible to human perception and cognition. I will be calling this landscape *digital-algorithmic politics*, and my concern will be the nature of the political agency that can be identified within it.

In this section, then, especially compared with what we saw in horizon 1 and will see in the conclusion, we will be less concerned with the ecological politics of this study's subtitle. We will not be looking particularly at agential alliances emerging from contexts we might broadly think of as biological, geological, or planetary, or at questions of environmental justice in the context of climate change. If this distinction can be said to carry much meaning, our focus will instead be primarily technological. But this distinction between the ecological and the technological of course carries little, if any, meaning. The imbrication of the digital-algorithmic processes with which we will be concerned and the biological, geological, and climatic developments characterizing the so-called Anthropocene is intensive and itself a defining feature of this time. The policing of illegalized climate migration we saw in horizon 1, for example, depends fundamentally on digital satellite images processed algorithmically in real time; the digital technologies of security and warfare we will encounter her are profoundly implicated in such population movements; and the massive planetary infrastructure of digital-algorithmic culture both depends on no less massive extractive industries and is itself a major contributor of climate-altering emissions. And, to come a step closer to the specific concerns of this section, barely any significant mobilization in the area of ecological politics can today avoid operating via the modalities of organization, decision-making, and communication that define digital-algorithmic politics. If we want to understand the kinds of political agency at work within contemporary ecological politics, then, we must confront the implications for such agency of those defining modalities. That is the task of this section.

The politics of digital algorithms is everyday news, mostly in terms of the governmental security protocols these algorithms infiltrate into daily life or the human biases they encode and elaborate. As Davide Panagia writes, "There is no doubt that algorithmic governance is an actually existing regime of rule in contemporary democratic life. From sentencing algorithms in criminal trials, to predictive policing algorithms, to sorting algorithms that generate targeted campaign advertising, to biometric algorithms that monitor border crossings, and even the algorithms that manage traffic systems and other essential flows—all these versions of algorithmic governance participate in democratic politics and they do so at a microscopic and macroscopic level" (2020:110). As Panagia argues, however, though most critical accounts of these processes remain at an essentially reformist level, full understanding of the modes of this algorithmic political participation requires conceptual engagement with algorithmic ontology. In Panagia's account, the key dimensions of this ontology are the digital algorithm's virtuality and its reliance on processes of negative feedback. Negative feedback gives the digital algorithm a probabilistic orientation toward the future; crucially, this dimension "automatizes purposeful behavior" and "governs future outcomes, or outputs, through the perpetual regulation and alteration of threshold values" (126). The virtuality of the digital algorithm, meanwhile, means that these processes are taking place in ways that bypass the fundamental representational ground of conventional human political activity and theorizing. It is not a matter, here, of representations of the world that might be critiqued according to their greater or lesser accuracy (2020:118–19).

Indeed, as we will see, strictly speaking, it is not a matter here of any phenomenon that might be amenable to representation by human consciousness. It is this combination of probabilistic futural orientation and nonrepresentational effectivity that gives the digital algorithm its particular pertinence to this study. For our purposes, the signal point in the algorithm's circuit of input-output is its transformation of the former into the latter, as interpretation shifts into projection and a future is indicated; the point, that is, where the algorithm makes its decisions.

The participation of this algorithmic decision-making in all manner of social and political activities has thrown dramatic light on the relationship between human and technical forms. When the latter are conceptualized as tools, this relationship is configured as mutual exteriority, whose inevitable outcome is instrumentalization in one or other direction: either human beings remaining masters of their tools or rebel tools turning on their masters. ("The machines are taking over!") In either direction, the only conceivable mode is domination; for the human beings imagining this relation, the only affective tonalities available are accordingly triumph or panic. (For recently successful examples of this panic with reference to digital technologies, see notably Bridle 2018; and Williams 2018.) As we have learned through

Stiegler's development of Simondon, however, the relationship in question is in fact one of coindividuation: far from mutual exteriority, human beings and technical forms emerge transductively together. In its ignorance of this mutual implication, the fantasy of exteriority actually underestimates the full force of the technologies in question, letting human agents imagine that we are understanding and manipulating these forms from the outside, however much lip service we pay to the idea that these objects also act on us. As Panagia argues, responses to algorithmic governance that remain at the reformist level miss this fundamental ontological point, imagining that "an individualist ideal of agency, freedom, and responsibility" can persist to guarantee a relation of instrumental exteriority between ontologically transcendent human beings and their more or less servile machines (2020:112). As Louise Amoore writes, "One does not need to look beyond the algorithm for an outside that is properly political and recognizably of ethics"; "the algorithm already presents itself as an ethicopolitical arrangement of values, assumptions, and propositions about the world" (6). Far from a scenario that would locate values exclusively on the side of transcendent human users, against their inscrutable but ultimately legible instruments, we have mutually formative and transformative allies, ranged, as we will see, along a decision-making agential front of differentially distributed capacities and values.

No adequate response here can forget the defining existence of what Göran Bolin and Jonas Andersson Schwarz call "the metricated mindset" (2015:10). The very thought processes through which we are deciding our behavior on the basis of our new understanding of technical forms are, if we have understood correctly, no longer conceivable as transcendent in relation to these forms. When Katherine Hayles writes, for example (in lines informed by her earlier work on the neuroscience of reading; see Hayles 2012), that digital devices "bring about neurological changes in the mindbodies of users, forming flexible assemblages that mutate as information is gathered, processed, communicated, stored, and used for additional learning that affects later interactions" (2017:119), this is a spiral of interaction with neither origin nor end, in which the interacting elements are continually formed and transformed by this relation.

This is not the Fall; as Stiegler in particular shows, there is no pretechnical degree zero of (human) consciousness relative to which we might measure some posttechnical distortion. The exteriority of user to tool no longer applies: we must take literally Tiziana Terranova's statement that "the digital automaton unfolds in networks consisting of electronic and nervous connections so that users themselves are cast as quasi-automatic relays of a ceaseless information flow" (2014:383). Indeed, we will have to understand that in this spiral, the role of user itself now switches between human and nonhuman beings. In Bratton's words, the user "could be a person, but it could also be a high-speed trading algorithm, it could be an animal, vegetable, mineral, driverless car, whatever you want it to be. We see most of the traffic on the Internet is already non-human" (Kuitenbrouwer 2015: para. 17). And as Bratton continues, this has serious consequences for political thought, imposing the need to "separate the idea of the agency and political rights and responsibilities of the user from those of the individual human organism, and no longer understand these as isomorphically mapping on one another" (para. 20).

This section will accordingly explore some of the implications of such digitally distributed agency for our understanding of politics. In doing so, it will in particular look to extend the notion of decisive action we have been pursuing in this study as key to any such understanding. So far, we have discovered that distributed agency is compatible with decisive action in the first sense (i.e., that makes a major difference to a state of affairs), and, thanks to Stiegler, we have developed a theory of the decision itself as processually immanent. One reason for now turning to the digital-algorithmic is that, as we will see (and as is only too well recognized in ubiquitous commentary), certain of its key processes, including decision-making, are beyond human perceptual and cognitive grasp. In this context, the politics of the processually immanent, distributed decision are thrown into exceptionally sharp focus. With machine-to-machine conversations, the fully automated creation of financial value, or adaptive machine learning, we humans encounter a realm we could at best describe as *quasi*-phenomenal; that is, as occupied by events and processes—including the formulation and actioning of decisions—that might be amenable to representation by human consciousness, but not in the time of their appearance. As this realm is also where significant parts of political operations are taking place, we are accordingly going to need a model of political agency that can incorporate such quasi-phenomenal sites of decision into its depiction of agential alliances and that can accommodate forms of political decision-making that are indifferent to standards of representational fidelity. Encountering digital-human alliances, we will find radically distributed structures in which the politically partisan intervention of the alliance—which might itself take the form of a decision—both significantly reconfigures its situation and is composed of local decisions made at positions and through processes that remain mutually opaque. With the quasi-phenomenal, distributed political decision, we will thus be able to understand distributed political agency not only as making a drastic difference but also as encompassing decisions in the strong sense: that is, as decisive in both senses of the term.

PHENOMENALLY UNPREDICTABLE

First, then, we turn to the phenomenality of politics in the time of the digital algorithm. Reflecting in *Specters of Marx* on the variability across different sociopolitical formations of the boundary between public and private, Jacques Derrida describes the medium in which this boundary establishes itself as "the medium of the media themselves (news, the press, tele-communications, techno-tele-discursivity, techno-tele-iconicity, that which in general assures and determines the *spacing* of public space, the very possibility of the *res publica* and the phenomenality of the political)" (2006:63). The "*spacing* of public space" is a question of rhythm, of articulation in space and time; in Derrida's description, this articulation regulates access to the realm of the political as a realm of *phenomena*, available to perception and cognition within the lifeworld of those concerned. Derrida's reflections here are informed by his dialogue with Stiegler, notably those recorded in the discussions published in *Echographies of Television* (Derrida and Stiegler 2002, first published in French in 1996). Furthering these discussions in the second volume of *Technics and Time*, also published in French, in 1996 (twenty years before notions of "fake news," "impression management," and so on became common currency), Stiegler suggests that the effective elimination of delay in "real-time" technologies—that is, the reduction of delay to a negligible or imperceptible level within the human sensorium—produces a "temporal rapture" that itself leads to an "inability to distinguish facts from fabricated facticality" (2009a:116).

In this state of affairs, "the contemporary politician is increasingly a manager of opinion." In addition, "Public life . . . is a function of real-time procedures, . . . the goal of which is audience-control through quantified knowledge of current trends" (2009a:122–23; translation modified). It might even be, he proposes, that politics in the strict sense can be articulated only over the long term; to the extent that all politics entails deliberation and decision, its rhythm is scored along technologies of relay and, indeed, delay. When that rhythm sees this temporal gap narrowed to the minimum of real-time circulation, the result is "the end of strictly political programs coordinated with coherent and integrated choices made over the long term as functions of ideas and collective actions, which are replaced by objectives and strategies of communication" (123).

Stiegler's distinction here between the "strictly" political and whatever it is that replaces it under a regime of "temporal rapture" is not especially helpful. The pre- or anti-deliberative practices that replace (or, better, overlay) his ideal order of debate and long-term projection are still plainly political in other terms (in Leninist terms, for example, as pure power play). And we might observe that contrary to an idealization of post-Enlightenment "deliberative democracy," these other dynamics have always also been at work within politics "in the strict sense," not least in their efforts to circumscribe by race, class, ability, and gender the population whose deliberations get to count. What we should take from Stiegler's analysis is not this distinction, then, but the key point that any political order is organized by the rhythm of its defining technologies of circulation. On this basis, Stiegler's analysis emerges as strikingly prescient, given the challenge posed recently to the post-Enlightenment understanding of the public sphere by the major transnational political deployment of digital-algorithmic technologies. Taking the form of unprecedentedly finely tuned communication with voters via social media profiling, and facilitated by the automated harvesting of data on a massive scale, this tactic looks to short-circuit this public sphere via a drive-based affective economy, itself sustained in part by the dispersed production lines of free attentional labor that we encountered in chapter 2.

As these features suggest, this challenge to the phenomenality of politics within the human lifeworld results primarily from two aspects of digital-algorithmic technologies: scale and speed. In terms of scale, the salient fact is what Galloway and Thacker call the "elemental" aspect of networks: namely, "that their dynamics operate at levels 'above' and 'below' the human subject" (2007:157). The technologies in question derive their power from their capacity to harness the massive to the granular: the step change they effect in taxonomic detailing is made possible by the massification and granular parsing of data sets. Relaying between the mega and the nano, these processes eclipse the Vitruvian scale: if the physiology of the human body acquires pertinence either as disassembled and dispersed within massive aggregates or as the locus on which their effects are registered, these effects are secured by the operation of the processes in question at levels out of all proportion to any individual human form.

In terms of speed, the relevant discrepancy is that between processes of electronic transmission in biological and information and communication technology systems. In 2014, the difference between the respective operational speeds of biological neurons and modern microprocessors stood at seven orders of magnitude (approximately 200 Hz vs. 2 GHz), while transmission via axons took approximately two million times longer than by optical fibers (Bostrom 2014:59, 272nn19 and n21). These speeds have mostly been discussed in relation to digital-algorithmic high-frequency trading (HFT), in which realm their acceleration has been particularly striking. In 2017, Hayles described HFT algorithms as conducting trades in "five milliseconds or less" (2017:131), and by 2019, Donald MacKenzie could quote data from the futures exchange Eurex giving the equivalent time as 84 nanoseconds (84 billionths of a second; 2019: para. 1). (For comparison, human consciousness takes about half a second to register perceptions [Hayles 2017:44].) As Hayles observes, this disparity radically estranges human and digital-algorithmic temporalities: the "microtemporalities" of HFT "are inherently inaccessible to human cognition, which can follow their electronic traces only in a temporal window that has, from the viewpoint of the algorithm, already faded into the past, but which for the inquiring human resides in the future of conscious recognition" (2017:144). Crucially for our purposes, Hayles points out that this "temporal gap . . . creates a realm of autonomy for technical agency" in which "algorithms draw inferences, analyze contexts, and make decisions in milliseconds" (142). Conversations between high-frequency trading algorithms produce (and wipe out) value in a realm beyond human perception and cognition, which register the consequences of some of these actions (this is, after all, their point, from the human perspective) but to which the vast majority remain opaque. "The players in these games are non-human," as Mark Coeckelbergh puts it (2015:108–9).

These effects of scale and speed place the operations of the digital-algorithmic "outside of the sensori-motor faculty that defines human experience" (Nony 2015:131); as Mark Hansen writes, "they seem to introduce levels of operationality that impact our experience without yielding any perceptual correlate" (2015: 4). In addition, non-rules-based adaptive machine learning and the development of accretive algorithmic ecosystems generate processes whose complexity is beyond explanatory reconstruction in the terms of human narrative and that, consequently, remain opaque to human agents (Danaher 2016:248, 254–55). This discontinuity gives rise to justified anxieties on the part of the human beings who find their capacities so outstripped, not least when it is instrumentalized for financial gain—as it is not only in HFT but already in everyday social media platforms, split between what Hansen calls "perceptually accessed media" and "media operating in the deep background" with "no direct experiential connection" between the two orders (2015:141, 73). In such cases, value is derived "by precisely not addressing consciousness directly" (Dieter and Gauthier 2019:75–76).

With human consciousness and perception partly bypassed, a financially profitable gulf opens between human and digital-algorithmic phenomenality. When digital-algorithmic processes become operational within the political realm (and further monetizable on the basis of this operationality), this gulf places in crisis the phenomenality of politics itself. The delay that Stiegler proposes as essential to deliberative democracy is indexed to the apprehension of phenomena within the human sensorium: when forms of political address, for example, are enacted, analyzed, and managed by procedures to which this sensorium has no access, politics starts to take place below and beyond the threshold of human experience.

Here, then, is a first dimension through which digital-algorithmic culture imposes itself as a matter of political concern: the specific phenomenality of those political operations enacted by massive computational capacity. At best, they might be apprehended—albeit as sublimely strange—by the objective phenomenology proposed by Hansen (who has gone furthest in modeling the weird phenomenality of technological processes that are constitutively opaque to human access). Like Ian Bogost's "alien phenomenology" (2012), Hansen's approach takes as its concern "the asubjectal subjectivity that correlates to the objective operationality of today's culture," in which "acts of sensing performed by machines . . . generate intentional or 'quasi-intentional' appearances of the world . . . that do not depend on consciousness or any other avatar of the subject"

(Hansen 2016:80, 86). An account of digital-algorithmic politics will somehow have to incorporate these "appearances" as the sites of meaningful decisive action that they also are, regardless of their "non-interpretability" by human actors (Danaher 2016:248). In this sense, digital-algorithmic politics will be defined by the opacity of the decision-making processes operative at one position in an alliance to users located and making decisions at another of its positions—that is, it will have to be understood as quasi-phenomenal.

A second dimension of equal importance to a coherent model of political agency within this culture shifts the register from the phenomenal to the governmental. This dimension has most notably been explored by Antoinette Rouvroy and Thomas Berns via the concept of "algorithmic governmentality." This concept—a major reference for Stiegler in *Automatic Society*—is defined by its authors as "the automated collection, aggregation and analysis of big data so as to model, anticipate and pre-emptively affect possible behaviours" (2013: para. 10). In their account, whereas statistical governmentality (as analyzed by Foucault and persisting alongside its algorithmic successor) entails reference to an "average" citizen, consumer, and so on, algorithmic governmentality entails "apparent individualization," "the idea of one becoming one's own profile, automatically attributed and evolving in real time" (para. 10). Crucially, for Rouvroy and Berns, and not least thanks to the discrepancy of scale we saw earlier, "algorithmic governmentality produces no subjectification, it circumvents and avoids reflexive human subjects, feeding on infra-individual data which are meaningless on their own, to build supra-individual models of behaviours or profiles without ever involving the individual, and without ever asking them to themselves describe what they are or what they could become" (para. 10). As a result, "The moment of reflexivity, critique and recalcitrance necessary for subjectification to form seems to constantly become more complicated or to be postponed" (para. 10).

The difference between algorithmic and statistical governmentality turns, for Rouvroy and Berns, on their respective relations to a posited average or norm. In their account, statistical governmentality interpellates individuals in relation to an abstract norm, thereby producing a space between such individuals and its own operation that maintains the possibility of dissensual critique. They argue that "datamining, used for profiling purposes, . . . following a correlation rationale to rebuild singular cases fragmented by coding, relates these singular cases not to a general norm, but only to a system of eminently evolving relations between various measurements that are not reducible to any average" (para. 2). (In this, they follow Foucault's analysis of the difference between discipline and security as modes of governmentality, with the latter now inflected algorithmically; see Foucault 2007:57; and on this, Amoore 2009:54.) Ironically, it is because "algorithmic governmentality thus focuses not on individuals, on subjects, but on relations" (Rouvroy and Berns 2013: para. 4) that its logic forecloses the relational space of contestation. Loyal to a Foucauldian tradition, Rouvroy and Berns insist that individuals are not simply in the grip of this mode of power. Rather, the significant point is that "irrespective of their capacity for understanding, willpower and expression, 'power' approaches them no longer on the basis of these capacities, but rather on that of their 'profiles' " (para. 14). Nevertheless, it remains their argument that "Algorithmic governance . . . neither produces nor provides an affordance for any active, consistent and reflexive statistical subject likely to lend it legitimacy or resist it" (para. 23). And so they ask, in conclusion, "Does the regime of digital truth (or digital behaviourism) not threaten, today, to undermine the very underpinnings of emancipation by eliminating notions of critique and of project?" (para. 45).

Similar in this respect to Stiegler's account of deliberative democracy, Rouvroy and Berns's description of the contestatory affordances of statistical governmentality clearly incorporates an amount of ahistorical idealism. Nonetheless, their model of algorithmic governmentality allows us to formulate a second essential question posed by digital-algorithmic political culture: what kind of political interpellation does this culture effect? Discussions of this question in political theory have tended to focus on the degree of fit between this culture and the forms of organization and representation that liberal democracies have inherited from the Enlightenment and the Industrial Revolution (whose articulating rhythm is, of course, that of print culture). Some on the left parse digital-algorithmic political culture through the existing categories of class mobilization: McKenzie Wark identifies a "hacker class" and a "vectoralist class," for example (Wark 2004), and Trebor Scholz proposes "Social media for worker mobilization" and "Platform cooperativism" as routes to a "People's Internet" in which "everybody can reap the fruits of their own labor" (2017:166, 174, 164). Similarly, Terranova demands that the profits generated by the monetization of data "be returned to those who actually produce them—that is, to living labor," and that "social networking platforms should be deprivatized" (2013:53), a position summed up in Evgeny Morozov's slogan: "*Socialize the means of feedback production!*" (2019:65).

Elsewhere, the mutation of the party form from the mass party of the industrial era, via "the cynically professionalised 'television party' [of] the post-Cold War era of high neoliberalism," to the digital or platform party is mapped by Paolo Gerbaudo (2019:14), whereas Zeynep Tufekci assesses the strengths and weaknesses of the digitally networked public sphere through analysis of contemporary social movements as networked protests (2017). Where Tufekci focuses on protest within the digital public sphere, Cass Sunstein, in turn, provides a broad assessment of the affordances of digital culture for the well-being of the public sphere in general, including the institutions of civil society that serve to maintain this sphere by guaranteeing discursive heterogeneity in public debate (Sunstein 2018). And Morozov's vision of collective ownership of the means of feedback production turns out to entail what we might call a digital civil society, in which "citizens could enlist allies and convince others of the virtues of their own readings of particular problems and proposed solutions to them," meaning that "deliberation-based democratic procedures could themselves be modes of problem-solving and means of social coordination," (2019:56), much as described, in fact, in David Karpf's account of current digital activism (Karpf 2017).

These political-theoretical approaches repeatedly suggest that the question of the kind of public convened by digital-algorithmic political culture requires consideration of the relation between this culture and its print-based industrial-Enlightenment predecessor. Currently, the two plainly coexist: it is less a matter of succession than of overlapping, or of what, borrowing from Raymond Williams, we might call the coexistence of *residual* and *emergent* forms (Williams 1977). Various commentators accordingly describe this relation as a kind of composite. For Tufekci, the success of networked protests depends on their ability to add to their model the organizational benefits of slower, predigital forms of civil mobilization (2017; and on this, see Cole 2019). And Ned Rossiter and Soenke Zehle note that even in a digital-algorithmic political culture thought to have short-circuited the articulation of liberal democratic political representation, "the concerns of a more conventional politics of representation are never far away, suggesting that the seriality of emergent political forms may translate across a representation/nonrepresentation divide" (2013:234). Rossiter and Zehle accordingly write of "the algorithmic constitution of new publics," defined precisely as "hybrid online/offline publics" (234, 239n33)—but just how radically should we understand this? In the accounts we have here been following from political theory and sociology, the composite form in question is invariably limited to the lamination of the layers on which human actors pursue their political activities. Action and medium are composite (online/offline), but the agents remain unitarily human and, indeed, unique. But persisting with this restricted understanding of composition risks seriously underestimating the questions posed by the quasi-phenomenality of digital-algorithmic political culture, which reshapes agency in a more significant sense.

For Rouvroy and Berns, this reshaping is the ultimate consequence of the elimination by algorithmic governmentality of the deviation that had remained possible under statistical governmentality. Whereas statistical reason conceptualizes deviation as its own necessary outside (with standard deviation as its immunological incorporation of an element of this outside), algorithmic reason in the age of machine learning conceptualizes "deviation" as a glitch, a stumbling block that can be accommodated by the system's capacity to adapt. "Algorithmic governmentality," they write, "with its perfect adaptation in 'real time' . . . renders the very notion of 'misfire' meaningless; in other words, a misfire cannot 'jeopardize' the system, it is immediately re-ingested to further refine behavioural models or profiles" (2013: para. 10). This integration of unforeseen deviant events into the self-improvement of a program matters for Rouvroy and Berns because the unforeseeable future is in their account the guarantee of an existential freedom coded as essentially human. It is, they claim, "the share of uncertainty, virtuality and radical potentiality that makes human processes free to *project themselves*, to relate themselves, to become subjects, to become individualized along trajectories that are relatively and relationally open" (para. 45).

The same point is made by Franco "Bifo" Berardi, for whom the "prescriptive inscription" at work in "the (self)construction of the automaton" "nullifies the (relative) autonomy of the future from the present" (2019: para. 14). Like Rouvroy and Berns, Berardi understands the stakes as those of human freedom: given this nullification, he asks, "do we still dwell in the sphere of human history? Or, on the contrary, have we transitioned into the sphere of the cognitive automaton, where history does not exist, where unpredictability is erased, and the infinite complexity of reality has been reduced to the combination of computable units?" (2013: para. 23).

From the humanist perspective, the confiscation of free human action by means of the incorporation of any futural bifurcation into the ongoing perfection of algorithmic programs would plainly put an end not only to history but to politics, understood as collective, partisan human action in the public sphere. But this perspective is doomed to either catastrophism or nostalgia—for the reason that it mistakenly conflates one form of transcendence with another. The drastic discrepancies of scale and speed between the human order and that of the digital clearly install the transcendence of the former by the latter. But this transcendence is perceptual and cognitive, it is not ontological. Only by confusing these forms of transcendence can Rouvroy and Berns and Berardi endow the digital-algorithmic with the power to cancel the futural projection that defines human freedom in the existential model. In this model, the human is ontologically transcendent in relation both to its tools and to time: through the mastery of the former secured by its exceptional cognitive capacity, it projects itself majestically beyond the present moment, thereby securing its freedom. Once these tools escape this mastery by transcending human perceptual and cognitive capacity (once the apprentice becomes the better sorcerer), this model has no option but to imagine the disastrous deflation of the ontological transcendence that had defined human freedom on the basis of this formerly transcendent capacity.

The shortcomings of this all-or-nothing vision give ample reason to refuse this perspective, however. Absent the fantasy of human ontological transcendence, the perceptual and cognitive transcendence of human faculties by digital-algorithmic processes announces not the end of politics but an emergent political form differently articulated. In the first place, of course, it is far from certain that the capacity of adaptive algorithms to learn from unforeseen events abolishes the future, any more than this dimension would be abolished by the ability of various biological organisms to learn from new experiences and to seek on this basis to manage their future existence. As Luciana Parisi writes, "the indeterminacies of programming . . . are able to unleash novelty in biological, physical and mathematical forms" (2019:13). Although Parisi's thinking is broader and more affirmative, we might return here to the example of HFT and in this respect cite the generation of black swan events (most notoriously, "flash crashes") by trading algorithms engaged in "continuous reciprocal causality through recursive feedback loops" (Hayles 2017:155). Drawing on Parisi's work, Patricia Clough accordingly argues that digital-algorithmic logic moves us "from an operational logic of closed systems and its statistically predictable populations to algorithmic architectures that override the possibilities of a closed system and predictable populations, opening . . . to the post-probabilistic" (2018:94–95).

An understanding of the digital-algorithmic milieu as characterized not solely by predictability but by "unpredictable consequences and emergent behaviors" (Hayles 2017:172), as well as an understanding of these consequences and behaviors as not only incorporated but also generated by adaptive algorithms, moves us from a disappointed humanist metaphysics to reflection on the conjoint action of digital-algorithmic and other actors in the elaboration of contemporary modes of existence. In this move, moreover, the unpredictable is reframed: no longer the exclusive guarantee of human freedom, it starts to emerge from algorithmic processes that in significant ways remain unavailable to human perception and cognition. (As Bratton writes, "it is their inhumanity that may make them most creative" [2015:81].) And the connection of the unpredictable to the notion of freedom suggests, rightly, that its fate is here indicative of a more significant redistribution of agency than can be grasped by this disappointed metaphysics. Beyond the hybridity or lamination of human activity within the online/offline public sphere, then, a model of digital-algorithmic political agency will have to engage composite alliances between human activity and the quasi-phenomenal operation of the digital as themselves local coalitions within broader agential fronts. We are now in a position to see what this might look like.

THE DISTRIBUTED DECISION

The challenge is well formulated by Adam Greenfield: "we need a better, more supple theory. . . . That theory needs to help us understand how agency and power are distributed across the meshing nodes and links of our collective being, how to evaluate the effect on our lives of that which cannot be understood in isolation and cannot be determined in advance, and how to assemble discrete components in ensembles capable of prevailing over the recalcitrance of things and actually making change" (2018:312). As Greenfield suggests, we need to start with the distribution of the agency that is operative within contemporary alliances across the multiple, mobile sites of their composition.

In their discussions of the existential, sociological, and political shifts introduced by the capillary penetration of digitally driven processes into human life, authors invariably respond to this distribution by noting the dispersal of the unitary human being that this is thought to entail. But even many of the best such authors (for example, Harcourt 2015; Koopman 2019; and Schüll 2016) focus their inquiry on the effects of such distribution for "our" (i.e., human) subjectivities—doubtless believing that existential, sociological, and political questions can be meaningfully discussed only when corralled in this way. The opacity of many digital processes to human representation leaves approaches founded on this restriction ill-equipped to delineate the distributions that are in fact operative, however. The effective alliances that compose this field remain unthinkable without the activities of Hansen's "asubjectal subjectivity" (2016:80), not as context, constraint, or prelude relative to uniquely human agency but as already participating in the constitution of an agential front. When Colin Koopman asks, "Who are we without all these identifiers, numbers, and other bits stored away in countless many data warehouses? Who could you be without your data points? What could you do?" (2019:6–8), we need to understand that the algorithms at work in these warehouses are also the locations of what "we" do, whether or not their processes are amenable to human phenomenal representation.

A focus on the human beings in this relation as generators of monetizable data—what earlier I called data farm animals—highlights the exploitation characteristic of the extractive data economy, in which (in McKenzie Wark's brilliant phrase) "one becomes at best a minor share-holder in one's own brain" (Wark n.d.: para. 16). If we recall from chapter 2 the coindividuation of human and technical forms, however, we might start to sense that the relation in question is more strangely structured than can be grasped by this focus alone. As a transductive relation, it involves terms that come into existence with the relation itself—meaning that the human beings exploited by data harvesting are already partly constituted by their relation to the technologies that effect this exploitation. This is not to say that such exploitation is either nonexistent or in some sense inevitable—we should recall with Stiegler that the relation in question might be differently inflected and is no more essentially toxic than it is essentially beneficial. What it does suggest, though, is that the resources we will need to think this relation, in the context of the agential alliances of digital-algorithmic politics, will not be found by continuing to index the production of value exclusively to the laboring human body and its organs.

When sociologist of digital labor Karen Gregory notes that "while 'you' may not ontologically 'be' a gadget such as a camera or a pedometer, as long as those elements are attached to you and producing trails of data, you are a generative source of data aggregation who becomes, often unknowingly, a constituent of data-based populations" (Gregory n.d.: para. 4), we should register her point that these populations are not composed exclusively of human beings and that their activities are not comprehensively available for human scrutiny. Beyond the influence of digital-algorithmic processes on human lives, then, beyond even the shaping or the monetization of these lives by these processes, we must include the processes in our model as participants. For reasons of scale, speed, or complexity, the processes that define some of these processes inflect their agential front in ways that the self-representations of human consciousness are unable to gather—but their effectivity and the interactions of this with other participants are irreducibly part of what we have to think.

If such coalitions of the human and the digital-algorithmic are of interest for our purposes in this study, it is in particular because of their implications for the question of decision-making. Recall that in horizon 1, I proposed that the assembled stakes of the confrontational action of an agential alliance define this action as decisive in the sense of transformatively significant. Via Stiegler, chapter 2 allowed us to conceive of the decision as immanent to the processual emergence of its location. The role of this section is accordingly to investigate the confrontational, distributed action of the agential alliance as decisive in the second sense: as comprising decisions that can be understood as decisions in the strong sense without recourse to the ontological transcendence granted by exceptional human capacities. Which is where the digital-algorithmic comes in. For existing alliances between human and digital participants are characterized above all by their distribution of the decision across their diverse and mutually opaque positions. And so the quasi-phenomenality of the digital-algorithmic takes on its full importance for our model: the decision—without which we cannot model political agency as I am seeking to do here—now emerges as itself composite, distributed, constituted from the contributions of human and nonhuman sites along the agential front. Moreover, as we have seen, and as we will now see in greater detail, in key contexts, both this composite decision at the level of the alliance and the local contributions of digital and human participants effect specifically political interventions. Reconfiguring the situation into which they intervene, they promote *this* tendency over *that*, advance *this* cause against its opponent, align themselves with partisan positions in a conflict over value and the composition of life in common.

Geopolitically, this distribution is most acutely evident in the area of digital security and warfare. Here, "advances in non-rules-based machine learning algorithms are producing new forms of political authority," in which "the algorithms authorize what or who is surfaced for the attention of a security analyst who, in turn, cannot meaningfully access this process of authorizing and surfacing" (Amoore and Raley 2017:6). This lack of access is crucial to the significance here of algorithmic decision-making: with machine-to-machine conversation and adaptive machine learning, and given the scale, speed, and complexity of the operations involved, the explanatory narrative of a given algorithmic decision is not necessarily available for human reconstruction. The example of this with the most extreme consequences would be so-called signature strikes, the military version of algorithmic governmentality in which targets are identified not as individuals but as profiles, "packages of information that become icons for killable bodies on the basis of behavior analysis and a logic of preemption" (Wilcox 2017: 16). In such instances, the drone pilot or security analyst is no longer the sole locus of decision, "for she is a composite figure of distributed human and non-human agency." Consequently, "the moment of security decision . . . remains an indefinite moment of profound uncertainty" (Amoore and Raley 2017:7), jointly taken by human and algorithmic actors with neither having straightforward—or, indeed, any—access to the others' decision-making processes. If we want to preserve some kind of human preeminence in the process, we might allocate different parts of the decision to different actors—with algorithms responsible for proposing targets, say, and humans for deciding their fate—but this will not achieve the desired end: if the decision to kill this individual might depend on that human being, the prior designation of the individual as killable clearly does not. The decision that "this individual is to be killed" is irreducibly and irrecoverably distributed.

As we saw in horizon 1, to insist in this way on the composite nature of an agential alliance is not to amalgamate all of its positions into an indifferent mass. On the contrary, such insistence allows what I have been calling *more and better* agential differentiation. Hayles's work on what she calls "cognitive assemblages" of human and technical participants, including in the case of drone strikes, is especially valuable in this regard. Exemplary in its detailed parsing of differentially distributed capacities, it emphasizes both that no position of decision-making within an alliance is reducible to any other and that no such position acts alone. In Hayles's account, an attack by a drone (unmanned aerial vehicle, or UAV) accordingly appears as follows: "Regarding ethical issues through the perspective of a cognitive assemblage foregrounds the interpretations, choices, and decisions that technical and human components make as information flows from the UAV's sensors, through choices performed by the UAV software, to interpretations that the sensor and vehicle pilots give to the transmitted data, on to the decision about whether to launch a missile, which involves the pilots, their tactical commander, and associated lawyers, on up to presidential advisors and staff" (2017:135–36). As Hayles writes, "Understanding the situation as a cognitive assemblage highlights this reality and foregrounds both the interplay between human and technical cognitions and the asymmetric distribution of ethical responsibility in whatever actions are finally taken" (136). Far from cloistering responsibility on one side of a gulf between disparate beings, this kind of heightened attention better delineates its actual differential distribution across sites of action.

On the basis of such more and better differentiation, and in addition to their geopolitical significance, the lessons of global securitization and drone warfare can thus also significantly inform a theoretical model of digital-algorithmic political agency. Through the sharpened account of responsibility imposed by their differential distribution of decision-making along fronts composed of human and nonhuman actors, it becomes possible to conceive of *decisive* agency as itself emergent and composite—and so, of such distributed agency in political form.

If we now turn toward policies of increasingly digitized contemporary states presented as addressing a more explicitly domestic realm, we can identify examples of this distributed and decisive political agency at work. First, we can continue to follow the crucial work of Louise Amoore and take the short step from securitization procedures enacted on a global scale to those enacted at the borders of and within the nation state (understanding that this distinction is solely heuristic; operationally, and as Amoore especially has shown, the two of course tend as closely as possible toward continuity). Here, too, we discover the conjoint decision-making activity of human and algorithmic analysts, with the difference that as we move away from explicitly military operations, the share of decisive action claimed by the latter begins to grow. In domestic security politics as what Foucault's reversal of Clausewitz calls the "continuation of war by other means" (2003:16; and see Amoore 2009:50), digital-algorithmic identification protocols "target individual bodies, designate communities as dangerous or risky, delineate safe zones from targeted locations, invoke the pre-emptive strike on the city streets" (Amoore 2009:50). With "the generation of probabilistic association rules" thus positioned as "the forefront of homeland security practices," algorithms define the security status of groups and individuals and so "appear to make it possible to translate *probable* associations between people or objects into *actionable* security decisions," in which the decisions in question are made by both human and digital actors (50–51). Just as the equivalent military software makes decisions on the killability of individuals upstream from the human drone operators to whom it delivers the resulting selection, so homeland security software "produces a screened geography of suspicion" in which "the identification of risk is assumed to be already present within the calculation" (52) and consequently distributes the decision on security status undecidably across its alliance. As Amoore puts it, "the border guard is somehow taken off the 'front line' " (52) in "a spatial and temporal *deferral of security decision*" (61) that is also its ultimately irrecoverable dispersal between distinct kinds of agents.

Moving further into the realm of explicitly domestic governance, we find a still greater share of decision-making processes claimed by digital-algorithmic operations. Philip Alston, United Nations special rapporteur on extreme poverty and human rights, consequently declared in October 2019 that "the era of digital governance is upon us" (Alston 2019:2), defined specifically by the automation of decisions in the areas of policing, security and surveillance, immigration, and welfare. Alston's remit sees him focus particularly on the arrival of what he calls "the 'digital welfare state' in many countries across the globe" (2). Looking ahead, he states that "there is little doubt that the future of welfare will be integrally linked to digitization and the application of AI" (19). This is the production of what Virginia Eubanks has termed the "digital poorhouse" (2018:12), in which "automated eligibility systems, ranking algorithms, and predictive risk models control which neighborhoods get policed, which families attain needed resources, who is short-listed for employment, and who is investigated for fraud" (3). The integration of digital-algorithmic systems into welfare provision is indeed well advanced: with reference to the United States, Eubanks notes that "automated eligibility is now standard practice in almost every state's public assistance office" (12).

In line with his remit, Alston is concerned to address "the dangers for human rights of various manifestations of digital technology and especially artificial intelligence," over and above their implications for "the traditional civil and political rights such as the right to privacy, non-discrimination, fair trial rights, and the right to freedom of expression and information" (Alston 2019:19). The crucial fact placing human rights in danger from the digitization of governance is, for Alston, the replacement of human decision makers by their digital-algorithmic counterparts, given what might at best be described as the indifference to such concerns on the part of the private-sector developers from whom public bodies are sourcing the software in question (11–12). Repeatedly, he emphasizes that these "new forms of governance . . . remove discretion from human decision-makers" (2), that they proceed "without the involvement of caseworkers and other human decision-makers" (8); those whose cases are handled in this way can, for example, find themselves stripped of benefits "automatically, without the involvement of a human decision-maker" (10). In Eubanks's summary: forty years ago, "human discretion still ruled the day"; today, "we have ceded much of that decision-making power to sophisticated machines" (2018:3).

Alston's attention is primarily directed to signal instances in which such approaches have generated injustice in order to underline the importance of human oversight. Emphasizing the discriminatory effects of these approaches in terms especially of race and class, and the self-validating algorithmic amplification of prior decisions, Eubanks goes beyond local instances to identify the systemic coherence of this tendency as "a feedback loop of injustice" (Eubanks 2018:7), "a generation of astonishing, sophisticated technologies that automate discrimination and deepen inequality" (204; and on the "racializing assemblages" produced by "more-than-human ontologies of digital architectures," see Dixon-Román 2016:489).

For our purposes, the salient fact is the inaccessibility of the digital-algorithmic decisions in question, which, even with such oversight, might prove irrecoverable. In the case of "the automation of eligibility decisions in Ontario in 2014 through the Social Assistance Management System (SAMS)," for example, the digital system "made decisions very difficult to understand" for both recipients and staff (Alston 2019:7–8). This opacity is frequently at the center of concerns about such automation. For example, reporting for Algorithm Watch on a case from the Swedish Public Employment Service in which 10–15 percent of automated decisions were "likely to have been incorrect," Tom Wills notes, "It is unclear whether it will be possible to identify and correct the erroneous decisions, and when exactly the problem started" (Wills 2019: paras. 1–3). From the perspective of the citizen who is "targeted by an algorithm," Eubanks writes that "you get a sense of a pattern in the digital noise, . . . but you can't put your finger on exactly what's amiss" (2018:5); more fundamentally, it also might be the case that no human analyst can trace the decision in question. Speed, scale, complexity, and non-rules-based machine learning sequester the digital-algorithmic decision for digital understanding, locking in the distribution of social policy decisions across the human/digital divide.

In warfare, security, and social policy, we can thus see the distributed decision in action—and the drastic existential stakes of the actions in question also serve to remind us of the confrontational stance of the alliances concerned. Sensors, software, operators, commanders, managers, government and other public officials: across these three areas, we see such participants composing agential alliances that operate decisively, in both senses of the word, on one side of what we might broadly term a clash of interests. These are political agential alliances, whose action is both distributed and decisive. In these cases, they are also aligned on the dominant side of existing power relations. Indeed, I have focused on these three examples precisely because of their often fatal and discriminatory outcomes: partly because this makes their confrontational character strikingly apparent, but partly also because the severity of these outcomes throws the harshest light on our understanding of the agential dynamics involved. Faced with such outcomes, the temptation to insist on the need for more extensive human oversight is strong; we should make our tools behave better. But the opacity inherent in these systems has already, by definition, withdrawn them from any such instrumental relation and, consequently, from any such comprehensive oversight.

This is emphatically not to argue for indifference or resignation in the face of these outcomes. Rather, it is to argue that effective engagement cannot be grounded in the valorization of a transparency that will always remain out of reach—condemning its adherents to the false alternative of passive acceptance, one the one hand, and Neo-Luddism on the other. (For the most extended account yet of an ethics and politics of digital algorithms that refuses fantasies of transparency and complete oversight in favor of starting from the "opacity, partiality, and illegibility of all forms of giving an account, human and algorithmic," see Amoore [2020:8].) Whether we are interested in addressing digitally enacted violence and discrimination or enlisting digital-algorithmic technologies in more progressive causes, we will have to understand these not as tools but as actual and potential allies.

In this sense, the quasi-phenomenality of composite digital-human decision-making has only provided particularly forceful evidence of the coconstitutive relation between human and technical forms—or, in Latourian rather than Stieglerian terms, of the heterogeneous alliance as the site of agency. And rather than Stiegler's model of politics as the decision on the mode of adoption of technical forms, Latour's notion of the alliance in fact proves especially helpful as a model in this case. Allies rarely, if ever, enjoy comprehensive insight into each other's decisions; the risk this imposes is part of the price paid by the allies for the enhanced effectivity offered by the alliance. Even if we want to think in exclusively human terms, the "other minds" problem makes this risk unavoidable. The mutual opacity that characterizes digital-human agential alliances is not, therefore, in itself anything either entirely new or inherently alarming, unless we conclude that the orders of magnitude that distinguish digital-algorithmic from biological cognition introduce a difference not only of degree but of kind. Even if this were the case, however, human actors are left with the same alternatives: passive acceptance, Neo-Luddism, or coalition. How, then, might we conceptualize the specifically progressive agential fronts to which such coalition might give rise?

We might get an intimation of this from the "weird solidarities" Gregory imagines at work in contemporary data alliances. As she writes, the value currently extracted from data "is already predicated on a social body and the generative connections that can be forged among its constituent elements"; and these elements, she specifies, "do not necessarily have to reduce to 'the human' " (Gregory, n.d.: para. 1). When Gregory writes that "the data economy extends and opens the human body to the pre-personal, to what we do not necessarily have conscious access to," for example, she has in mind the genetics by which we are traversed. But she might equally well be describing the impersonal memory that, for Stiegler, is embodied in technical forms. Crucial here is Gregory's insistence that we go beyond the dystopian visions called up by our disappointed metaphysics: that "we must also take up the notion of 'weird solidarity' to explore how the 'more-than-us-but-not-us' is giving rise to new forms of sociality and social relations" (para. 6). In Gregory's words, "A map of this weird terrain is necessary to foster a politics of solidarity that understands how and where value is being produced" (para. 7). Just as Marx identifies the conditions of possibility of revolutionary action precisely in the industrial concentration of the working class, so Gregory allows us to think that the already existing fact of intimate, saturated coexistence between human and digital technical forms opens new forms of political action. We might describe this as composite agency reinflected as a solidarity through which to contest the political-commercial reduction of its defining relationality to mutual instrumentalization. Gregory writes, "A weird solidarity is already being created among us, in and through the ubiquitous project of building algorithms into every facet of day-to-day life. That solidarity is an essential aspect of the aggregation of the data. That solidarity is there, and therefore it is there for us to see, to experiment with, and to build from" (para. 9).

If Gregory mostly inflects her weird solidarities in the direction of more attentive modes of coexistence, I want to expand her proposal by drawing out another aspect, implied but not developed in the phrases I have been quoting and toward which I have been moving here; namely, that these new solidarities might also be fostering politically progressive forms of distributed digital agency. We tend to think about the digitization of politics in two ways, focusing, respectively, on new modes of collective action; deploying new tools; occupying new, hybrid online/offline spaces; constituting a hybrid public sphere, or on the new forms of address and assembly that gather constituencies through user profiling. But across and beyond these examples, we see the newly distributed form of the composite digital-human decision.

An understanding of digital-algorithmic political agency cannot content itself with describing the new affordances of our digital tools, as if the distributed decision-making characteristic of digital-human alliances had no effect on the nature of the agency concerned. The distributed structures defining what Bennett and Segerberg term "connective action" (2015) distinguish it essentially from more centripetally structured collective action. Although the two clearly coexist in most contemporary activism, the former sees the decisions that characterize its activity spread along its agential fronts, at human and digital sites that are not mutually transparent.

The "weird solidarity" imagined by Gregory is weird for this very reason: the quasi-phenomenal nature of its constitutive decisions for its various participants. From the hacktivism of Anonymous to the clicktivism of 38 Degrees, from #MeToo to #BlackLivesMatter to #RhodesMustFall, varied and antagonistic forms of interventionist digital solidarity are already in evidence, including opposition to the fatal and discriminatory actions of dominant digital-human alliances in such areas as securitization, policing, welfare, and voter registration. The agency displayed by these alliances constitutes a significant development in political culture, presenting new kinds of antagonistic intervention that crystallize where the exercise of power meets the dynamics of group formation and the modalities of action operated by their platforms. These embed digital-human decision-making processes within the confrontational agential alliance as a whole, which also comprises other participants such as funding mechanisms, media outlets (print, broadcast, and digital), and the forms of a predigital civil society, including street protests, speaker meetings, and legal action. And the action of these alliances takes place right along this front, with particular militant inflections at key positions. But the alliances in question are irreducibly defined by the visibility and anonymity, the speed and social penetration, the geographical concentration and dispersal, and the deliberative plasticity of their particular digital-human coalitions. That is, they are defined by the particular mesh they present of human and digital-algorithmic decision-making, which at their respective positions along the agential front make local decisions that define these and other qualities of the alliance in action and thereby compose its partisan political alignment.

This is how the quasi-phenomenal character of the digital positions its platforms and programs, not as tools manipulated by human users (despite the persistence of that term) but as participants in alliances. When decision-making is distributed across human-digital chains whose positions are mutually opaque, the process in question can no longer meaningfully be understood as determined by transcendent human intention. As the ubiquitous metaphor of "black-boxing" insists, a tool that is no longer transparent to its user is no longer a tool. If the fantasy of instrumentalizing transcendence always evaded the cooperation at work in what it called "tool use"—as Bateson would remind us, neither the human arm nor the axe alone has much hope of chopping down the tree—the quasi-phenomenality of digital-human decision-making renders that fantasy definitively inoperative. What is more, the capacities of the participants in such processes are themselves defined by their differential distribution across the alliance. As we have seen, human decision-making, for example, is here characterized by its functional relation to and difference from the digital within the actions of its alliance; and what Panagia calls the "dispositional powers" of digital algorithms—their capacity to reconfigure "relations of peoples and things in space and time" (2020:110)—are similarly both defined and enabled by their differential relation to the capacities of other participants, including human. And crucially, if the actions of such alliances are politically decisive, interruptive, and—in some cases—emancipatory, the mutating "real-time" rhythm of these actions sees authorship of even key moments similarly distributed across the alliance, as its decisions emerge in part via quasi-phenomenal operations of data processing, feedback, anticipation, and modification.

To pluralize authorship of political action in this way is not to undermine the contributions of the human participants concerned. On the contrary, in identifying eminent sites of accountability, it offers *more and better differentiation* of how and where effective interventions are made how and where the contributions of human participants interact with those of their digital allies, as well as what kinds and degrees of responsibility might be attributed to these different participants, including in actions that have emancipation as their finality. In response to such pluralization, there is no need to be nostalgic for a golden age of deliberative democracy (in the agora or the town hall) before human speech fell from its preeminence as the exclusive medium of politics. If such earlier versions were themselves also organized by the confluence of economics and technics (and most often in discriminatory ways that the pseudo-universalist pull of appeals to human speech can only obscure), conversely, these more recent forms plainly sustain spaces and practices of deliberation and contestation. Some of the forms in question certainly function to short-circuit such practices; the alliances of social media platforms, quasi-legal funding mechanisms, fine-grained user profiling, and populist grotesques that have characterized so much recent "digital politics" are plainly anti-deliberative in orientation. And others, such as the discriminatory forms of digital biopolitics we saw earlier in the analyses of Virginia Eubanks or Philip Alston, no less plainly constitute regimes of social exclusion that inscribe in newly capillary ways the fundamental police distinction between those who do and those who do not count as members of the demos. But these exclusionary and repressive effects are not in themselves attributable to the digital-human character of the alliances concerned, as is readily indicated by the existence of those digital-human alliances whose political interventions are contestatory and—potentially—emancipatory, from Minneapolis to Johannesburg to Cairo and beyond.

The point of this section, then, has not been to evaluate the forms of digital politics as ultimately reactionary or progressive. It has, rather, been to identify the specific political significance of the distributed mode of digital-human decision-making. The first part of this significance is methodological and descriptive and works in the direction of the more and better differentiation I have been emphasizing through this study so far as a key consequence of understanding political agency in distributed terms. If the rhythm and composition of the practices that make up digital politics can no longer be grasped whole by human perceptual and cognitive faculties, this itself can produce a finer-grained exploration of political agency as the differentially—indeed, often discriminatorily and exploitatively—configured meeting place of the socioeconomic, the existential-affective, and the technical. Such exploration is ongoing, as we saw earlier in the work of Louise Amoore and Karen Gregory, for example, alongside that of Eubanks and Alston.

The second part of the political significance of distributed digital-human decision-making is conceptual and analytical and goes to the heart of my aims in this study. Through this section, we have come to understand digital-human political agents as composite alliances engaged in confrontational struggles that are decisive in both senses of the word. Turned probabilistically toward the future, and operating without the human sensorium, digital-algorithmic participants in politically effective alliances make local decisions that mesh with those of their human allies to compose action with drastic consequences. And so we now have a picture of political agency as both distributed and decisive in the full sense: of plural alliances not only making a drastic difference (as was the case by the end of horizon 1) but also formulating and actioning decisions at the various sites of their composite agential front, which thereby enacts a composite partisan commitment within a conflict over value.

Such, then, would be politics as understood from the perspective of decisive distributed agency. If the agora (or the town hall) and the demos clearly still feature, as we have seen, they do so not as an ideally transparent forum and the constituent power that brings its claims to discursive representation within this luminous space but, rather, as a space and as participants within a greatly expanded field, traversed by agential fronts along which clash ontologically diverse alliances whose politically decisive actions are composed of local contributions defined in part by their mutual opacity. This is in fact a politics that can no longer imagine itself to be grounded solely, or even primarily, in a space defined as fundamentally representational—indeed, the digital-human political alliance is the indicator par excellence of the expansion of the political field beyond representationality that is operated by a model of political agency as ontologically distributed. In this sense, the quasi-phenomenality of digital-human political agency not only requires a politics in which representation plays a part but is not the whole story. This quasi-phenomenality might also be understood as inviting a more sharply critical account of forms of actually effective political action, precisely thanks to its unavoidable and unavoidably effective relativization of representation as the ground of politics. For there is no need to panic, in the face of these patches of opacity; the transparency of the space of political representation was always an idealist, exceptionalist fantasy. As we learned especially from Jane Bennett and Karen Barad in the introduction: parsing agency as distributed "requires that much more attentiveness to existing power asymmetries" (Barad 1997:219). If these partisan clashes of digital-human alliances look unlike a conventional picture of politics, one virtue of this change of perspective might perhaps be that the differential quality of this composition—and so also the power differentials, discrimination, and exploitation that form part of the distribution of agential capacities within and between conflicting alliances—can no longer be ignored.

* * *

With its theorization of the decisive distributed political agency, this picture accordingly ushers in the full model I am seeking to develop here. In chapter 2, we saw Stiegler formulate the decision in the mode of reflexivity without transcendence; with political agents as alliances of differentially distributed human and nonhuman decision-making processes, we have now been able to model the political decision as fully immanent to the emergence of its site from within the array of beings defining its situation. We can now understand distributed political agency as decisive in both senses—and, thus, politics itself as an activity of composite alliances of ontologically diverse participants. For the possibility of seeing such alliances as zones of decision-making crucially develops our account of distributed political agency.

Thanks to the evaluative and selective aspects of such decision-making, we can see how distributed political agency can open to a normative concern for the future without any need to ground this concern in human-exceptionalist claims to monopolize a capacity for projection. And if we recall Oliver Marchart's defining conditions of politics (2018: 36; as discussed in the introduction), we can say that in addition to the qualities of collectivity, conflictuality, and partisanship in evidence in our model of distributed political agency by the end of horizon 1, this opening to the future now allows us to identify qualities of organization and strategy. With these qualities, which the alliances we have considered in this section certainly exhibit, our model of distributed political agency starts to make good on the normative commitment to the future we owe to Stiegler's account of the decision.

Our remaining task is accordingly to explore this future: specifically, the future as the dimension of political prescription and commitment. We have seen in this section that the exclusive humanist claim on the future as guarantee of freedom is without foundation. What we have not yet seen in detail, though, is how the futural commitment essential to a full account of political agency can be understood in processually immanent terms. Chapter 3 is charged with demonstrating this possibility.

3

CATHERINE MALABOU

"There Is Nothing Beforehand"

Bypassing the ontological transcendence threatened by reflexive intentionality in its human-exceptionalist mode, the quasi-phenomenal political decisions taken across antagonistic alliances featuring digital and human participants build on Stiegler's account of the immanent recursive decision as the polemical affirmation of a value. This enables us to understand distributed political decision-making as immanent to the emergence of its composite site within its extended situation. In the elaboration of our model, we have accordingly reached the point at which the intervention of a heterogeneous agential alliance might be understood as decisive in both senses of the term: as making a major difference to its situation and as the work of a decision, without this decision resurrecting its location as transcendent in relation to this situation. The generalized quasi-phenomenality of the composite, distributed political decision—the fact that the respective contributions of the various participants remain opaque to one another—sets a further challenge, however. How can we understand such a decision as an act of commitment and responsibility?

If Stiegler gave us the immanent recursive decision as reflexivity without transcendence and transvaluative futural commitment, the introduction of quasi-phenomenal distributed decision-making complicates matters by multiplying the sites of the composite decision along an ontologically heterogeneous agential front. Stiegler showed us that decisions are never made by human beings alone; what we now need is a mechanism by which to understand the futural commitment enacted in such decisions as itself distributed across an agential alliance. Our guiding commitments in this study—like Stiegler's, moreover—rule out bringing such commitment into the all-revealing light of solely human understanding. Not only is this fantasy transparency designed to ground the transcendence of human reason in relation to its milieu, it is also, by definition, deluded, mistaking its particular light for a universal solvent. The quasi-phenomenal composite decision will not be justified by being retrospectively assimilated to an expanded or enhanced realm of human possibility. Rather, we will need a way to describe the emergence of a futurally committed, responsible decision that can secure these specific qualities against the gravitational pull of human ontological transcendence. The overall claim of the following chapter is that we can develop such a description from the work of Catherine Malabou.

Like Stiegler, Malabou features here as offering material we can use in the ongoing construction of our model. Like chapter 2, therefore, this chapter is interested less in exposition than in deployment. To identify this material, however, we will again need to suspend the vocabulary through which we are building this model for most of the chapter. We will spend time in Malabou's world, exploring her guiding motif of plasticity and the problematic it governs, and derive from these the structure by which we will account for the composite decision as a site of future-oriented commitment and responsibility without compromising its distributed character.

If Malabou's work will thus provide a key element in our ongoing elaboration of a model of distributed and decisive political agency, this element will not be derived directly from Malabou's own extrapolations of the politics of her core concept of plasticity (such as the account of the politics of neuroplasticity in what remains her best-known work, *What Should We Do With Our Brain?*). I will certainly be tracing some of these—less for their determinate content, however, than for their further clarification of the structure of agency Malabou's plasticity allows us to think. For this concept serves principally to articulate an account of drastic change as continually possible and processually immanent; indeed, in its name, she firmly rejects any appeal to a transcendent outside as the source of such change. Her philosophy thus offers, as Nancy D. Nisbet writes, "a remarkable philosophical apparatus that offers support for the emergence of immanent change" (2018: n.p.). Moreover, this insistence on the possibility of change as immanent strong discontinuity is habitually signaled in her writing by the figure of the *accident*—of which she is consequently a major thinker. For our purposes, it is of crucial importance that Malabou understands the accidental emergence of a locus of decision as a responsible commitment to the future. In her monograph on Hegel, she describes this emergence in the process she terms "plastic reading." Malabou's plastic reader emerges accidentally *in and as* this decisive location; and, importantly, this emergent plastic reader remains unspecified, allowing the futural commitment marked in its emergence as the site of a decisive interpretation to float free of any attribution to the categorically fixed capacities of given types of being.

This, then, is the structure that will allow us to reconcile the quasi-phenomenality of the composite decision with the future-oriented commitment and responsibility required by a meaningful model of effective political agency. The transvaluation enacted in Stiegler's quasi-causal, recursively immanent decision on adoption already opened us to the future, in its partisan commitment to promote *this* tendency, *this* value over that. The probabilistic decisions of the digital algorithm maintained this futural orientation while reuniting it with our model of distributed political agency. What Malabou now offers is a specific account of the structure by which this futural commitment to a value can remain immanent to its accidental emergence in a moment of decisive change.

In order to derive this structure from Malabou's work, we will need first to work through the relevant aspects of her account of plasticity. In particular, I will consider her definition of plasticity in terms of the capacity to give and receive form (including in the mode of explosive deformation), her related insistence on thinking change without reference to transcendence, and her configuration of the accident as immanent strong discontinuity. After some consideration of what Malabou presents as the "politics of plasticity," I will offer my unpacking and development of her concept of plastic reading, which will constitute her major contribution to the construction of the model I am building throughout this study.

As a key thinker of the accidental and the automatic, Malabou has often been accused of offering nothing to a meaningful understanding of political agency—accused, that is, by those for whom such an understanding mandates the exclusion or neutralization of these awkward aspects. We will consider such criticisms in due course. For those of us seeking an understanding free from the human exceptionalism by which these criticisms are habitually underpinned, however, Malabou's commitment to the immanent processuality signaled by such motifs makes her a vital interlocutor.

GIVE, RECEIVE, EXPLODE

When she started thinking about plasticity, Malabou says, "I glimpsed the appearance of something amounting to a new motor scheme, the accomplice of a new era" (2010:17). The concept and figure of plasticity run through her work from the start, according to a consistent characterization. In the introduction to *What Should We Do With Our Brain?*, she provides a helpful summary of her understanding of the term, which is organized into two aspects. The first is its core meaning; namely, the mutability of form, which itself divides into two: form can be both given and received. "According to its etymology—from the Greek *plassein*, to mold," Malabou writes, "the word *plasticity* has two basic senses: it means at once the capacity to *receive form* (clay is called 'plastic,' for example) and the capacity to *give form* (as in the plastic arts or plastic surgery)" (2008b:5). Sculptors and plastic surgeons shape forms; as plastic materials, wood, marble, and concrete are amenable to being shaped. (Hence the most familiar meaning of "plastic" in everyday use, to which, as Alexander Hope notes [2014:334], Malabou oddly devotes little attention, observing only that "*Plastic material* is a synthetic material which can take on different shapes and properties according to the functions intended" [2005:9].)

The second aspect by which Malabou characterizes plasticity is more surprising, at least for a Anglophone audience: "it must be remarked," she adds, "that plasticity is also the capacity to annihilate the very form it is able to receive or create" (2008b:5). In Malabou's French, "*le plastic*" is the form into which the English phrase *plastic explosive* morphed as it entered the French language (in 1943), and the derived terms *le plastiquage* and *plastiquer* are everyday words for "bombing" and "to bomb" using plastic explosives or similar material. There appears to be an obvious slippage at work here, in which an adjective has acquired the properties of the noun that it qualifies; the explosiveness of Semtex, say, would seem to have nothing to do with its plasticity. As Tom Sparrow writes, "while plastic explosives are, of course, designed to blow things up, this does not mean that explosiveness is implicated in the concept of plasticity itself" (2014:227–28). On the other hand, the nature of the damage plastic explosives can cause has everything to do with the composition of the material, which brings not only increased portability and newly precise positioning but also increased velocity of detonation. Whether or not we choose to grant Malabou her association with *le plastic*, and although in this it is hardly unique, we can at least agree with her that as a term, *plasticity* is itself plastic, taking on more than one form. As she puts it in *The Future of Hegel*, "The plasticity of the word itself draws it to extremes, both to those concrete shapes in which form is crystallized (sculpture) and to the annihilation of all form (the bomb)" (2005:9).

This destructive side to plasticity is stressed by Malabou, who rightly notes that notions of *formation* prefer to emphasize the positive. "In science, medicine, art, and education, the connotations of the term 'plasticity' are always positive," she writes. "No one thinks spontaneously about a plastic art of destruction. Yet destruction too is formative" (2015b:3–4). We might be happy to acknowledge a limited degree of negativity in the work of "plastic construction," but we resist the thought that explosive destruction is also a *change of form*, with its own phenomenology: "Something *shows itself* when there is damage, a cut, something to which normal, creative plasticity gives neither access nor body. . . . A vital hitch, a threatening detour that opens up another pathway, one that is unexpected, unpredictable, dark" (2015b:6).

Malabou's principle exploration of the phenomenology of destructive plasticity comes in *The New Wounded*, with its reflections on the significance of the brain damage experienced by trauma sufferers, Alzheimer's patients, and the victims of severe cerebral lesions. Throughout this exploration, she is especially interested in the duality of destructive plasticity; as she puts it, "Such plasticity—herein resides its paradox—ultimately remains an aventure of form" (2012b:17). In these cases, we are certainly "far from the sculptural paradigm of 'beautiful form' " (17): there is nothing that might be celebrated here as happily creative. And yet, the destructive change in question gives rise to the appearance of a new form. Accordingly, "If brain damage creates a new identity, this can only be *creation through the destruction of form*" (17). We are faced with the difficult reality of transformation through destruction, which Malabou seeks to capture in her paradoxical formulations: this is "a strange sculptural power that produces form through the annihilation of form" (49). In Malabou's interpretation of these cases, "It is entirely possible that there will be *no relation* between the identity that comes before a lesion or trauma and the identity that comes after" (19). Across this destructive change, we witness a complex regime of continuity in and as rupture, "a discontinuous process" (152), or what I will here call "dis-continuity."

That Malabou principally examines the work of destructive plasticity in the context of irreparable injuries to the brain eloquently indicates just how somber this other path can be. This is change as irreversible, irreparable, irredeemable loss, an abyss through which transition becomes a shattering substitution inassimilable to all stories, however supposedly bracing. In Jairus Grove's stunning formulation, "The thing that changes itself dwells at the precipice of nonbeing" (2015:258). The irreducibility of this dimension means that Malabou's concept is not some facile celebration of irreducible freedom or creativity. It is perfectly possible, Malabou points out, to understand plasticity in reductively upbeat terms, as if it were only ever positive (2009:217). The insidious ideological version is the neoliberal promotion of flexibility as a virtue, against which Malabou explicitly sets her concept of plasticity throughout *What Should We Do With Our Brain?*

It is not only in its explosive, destructive dimension that plasticity demands to be thought as more than just happily creative, however. If neuronal plasticity is what makes habit possible, for example, then it is also what enables the development and imposition of unthinking obedience. Positioning herself against the boosterish insistence on every change as a welcome opportunity, Malabou here follows Bourdieu's reading of Pascal to point out that the physical internalization of social discipline, for example, which permits the exercise of social power, is itself permitted by the neuroplasticity that allows the acquisition of new habits. "While it appears to be the sign of biological indeterminacy," then, brain plasticity "also serves to legitimize new ways of standardizing psychosomatic expressions" (2019a:98). What is more, for all that its destructive dimension exceeds the limited regime of construction as generally understood, plasticity as Malabou uses it does not indicate unlimited transformability. Mutability is irreducible, but any one plastic form is not infinitely mutable: "the adjective 'plastic,' while certainly in opposition to 'rigid,' 'fixed,' and 'ossified,' is not to be confused with 'polymorphous.' Things that are plastic preserve their shape, as does the marble in a statue: once given a configuration, it is unable to recover its initial form. 'Plastic,' thus, designates those things that lend themselves to being formed while *resisting* deformation" (2005:9). It is precisely because it resists deformation that plastic form can explode: there are pressures that it cannot accommodate. Plastic form is finite, and its destructive dimension signals this.

For Malabou, then, plasticity is "this place of contrasts where form forms itself and at the same time deforms itself, where it acquires consistency and explodes like a bomb" (2005:187, translation modified): "Between the emergence and the annihilation of form, plasticity carries, as its own possibility, self-engendering and self-destruction" (2005:193). Within this duality, and for all her warnings about the facile insistence on all change as positively creative, Malabou does, in fact, want to hold onto something perhaps minimally positive in destructive plasticity. The stakes here will turn on just what kind of creativity we are talking about. In *The Future of Hegel*, Malabou refers to "the chance . . . that the *explosive side of subjectivity represents*" (2005:187; translation modified); if it is not always positive, plasticity does always offer possibility: the possibility that something will happen, not necessarily something welcome or, indeed, assimilable in any way. But some kind of change, at least. As we will see, this possibility is crystallized for Malabou in the figure of the accident. Before we can explore her use of this figure, however (which will be crucial to her importance for this study), we must first further unpack her understanding of change.

NOWHERE TO RUN

The first thing to note about change as Malabou understands it is that it comes from nowhere other than right here. In Malabou's conception, plasticity is a way of thinking change without having to explain this by reference to an outside, to some other scene. In this sense, the accident is the very figure of strong emergence—of strong discontinuity that remains immanent to the system within which it appears. As she puts it in *The Heidegger Change*: "Modification can only . . . intervene *within the world*; where, that is, there is *neither inside nor outside*. . . . There is neither interior nor exterior; there is just the world, which is more intimate than any inside and more alien than any outside" (2011a:219).

Using Heidegger's own idiom (if not exactly sticking to his understanding of "world"), she later develops this thought further: "Being-in-the-world is the experience of an absence of the outside. Everything that happens, then, only can happen right in the world, and this mode of transpiring 'right in' brings us back to the heart of the movement of modification" (229; translation modified).

So at the heart of change is the fact of its unqualified immanence. "There is neither inside nor outside," writes Malabou (now in Deleuzian vein), "just levels of intensity" (2011a:238). In this model, specifically destructive plasticity represents what happens to flight where no flight is possible—what happens to escape when there is no outside: "metamorphosis by destruction is not the same as flight: it is rather the form of the impossibility of fleeing. The impossibility of flight where flight presents the only possible solution. We must allow for the impossibility of flight in situations in which an extreme tension, a pain or malaise push towards an outside that does not exist" (2015b:10). What Malabou calls an "ersatz of flight" (2015b:18) appears as an answer to the questions, "What is a 'way out'; what could be a 'way out' when there is no outside, no 'elsewhere'?" To understand such an answer, we have to grasp that "it is not a question of how to escape closure but rather of how to escape within closure itself" (2010:65). Malabou concludes, "I name 'plasticity' the logic and the economy of such a formation: the movement of the constitution of an exit, precisely where no exit is possible" (2010:66; translation modified).

The conceptual density of such formulations denotes the seriousness with which Malabou takes change as destructive plasticity. As her language of flight suggests, Malabou is here pushing on a Deleuzo-Guattarian affirmation of the ineliminable possibility of a line of flight, immanent to its territory (see Deleuze and Guattari 2014), and what this reveals is the importance of negativity in her thinking. The line of flight inscribes the virtual resources of the preindividual milieu within any individuated form; as no process of individuation exhausts these resources, any form is always open to the possibility of change, displacement, or—in language closer to Malabou's—transformation. The pressure Malabou brings to bear on this scenario argues that it remains necessary to think not only creative transformation along a line of flight but transformation as destruction of form, as deformation. Of course, Deleuze-Guattari understand perfectly well that transformation can occur as destruction, but for them, this is always a risk, the extreme danger of a negativity—"catastrophe," "demented or suicidal collapse" (2014:178) that requires the formulation of appropriately nuanced, cautious displacements. It is because destruction is always a risk that creativity must be resourceful. The axiomatic difference between Deleuze-Guattari and Malabou here concerns this relation of destruction and resources. For Deleuze-Guattari, there are always resources ("have a small plot of new land at all times," they write [2014:178]), and so destruction is a risk to be avoided in the pursuit of creative transformation. For Malabou, sometimes there are no resources—at which times, destruction is the only transformation there is.

In what Malabou accordingly calls "the continuous implosion of form" (2010:57), plasticity indicates that any happening at all, any random occurrence—what we might think of as the simple fact of time, however closed the system ("I have always associated plasticity with time," says Malabou [2019b:241; and see Deleuze 1989:77–78])—can *by accident* come to be an instance of drastic change: "The possibility for a closed system to welcome new phenomena, all the while transforming itself, is what appears as plasticity. Here again we find the process by which a contingent event, or accident, touches on the heart of the system, and, in the same breath, changes itself into one of the system's essential elements" (2005:193). For Malabou, plasticity thus marks the irreducible possibility of change without its having to be guaranteed by the transcendence of an outside: "Plasticity designates the form of a world without any exteriority," as she puts it (2010:67). This is why Malabou's apparently paradoxical formulations—"the constitution of an exit, precisely where no exit is possible," and so on—are not "mere" paradoxes or pirouettes. They might appear so—but only so long as we insist on indexing change to some ultimately redemptive other space, some outside or elsewhere. For Malabou, by contrast, it is just because there is no elsewhere that change is possible here. Nonetheless, "the form of a world without any exteriority" is, to say the least, a pretty chilling phrase in a global context saturated by the dogma of marketization and the monetization of the neurological. And this context does indeed form the horizon of Malabou's reflections here, as we will see when we return to Malabou's interpretation along these lines of the geopolitics of globalization. But to her mind, crucially, "a world without any exteriority" designates precisely the possibility of challenging this context, inasmuch as—surprising though this might at first appear—the refusal of exteriority is the first step to understanding the irreducible possibility of change in a context from which any such possibility has supposedly been evacuated. And across her thought in general, this refusal of exteriority is articulated primarily in terms that cannot but underline her relevance for our purposes in this study: namely, as a refusal of transcendence. On which she is, moreover, categorical: as she says in an interview, "I don't believe in transcendence at all" (2008a:10).

According to Malabou, it was the study of brain plasticity that led her to understand this way of thinking about change; what she calls "mutability in a moving whole stripped of transcendence" (2012a:265–66; my translation). In this understanding, the unqualified immanence of plastic change relieves us of any need to ground the possibility of change in some external instance by which the being undergoing such change would always and everywhere be surpassed. Plasticity, that is, means no transcendence and so no *other*. "To put it differently," she writes, "plasticity renders possible the appearance or formation of alterity where the other is absent. Plasticity is the form of alterity without transcendence" (2010:66). This refusal of alterity as transcendence seems no less chilling than the vision of a world with no outside: what kind of murderous solipsism is this? But Malabou is well aware of her readers' likely reaction:

> This transformation, I will doubtless be told, is a reduction of the alterity of the other, of he or she who is loved and their foreignness. I believe, however, that there is nothing to this: the alterity of the other is perhaps just what announces itself in and for me, and as this very strange capacity I have to metamorphose, displace myself, and to be born beyond myself so as to become akin to the other—a capacity whose existence the other reveals to me. (2011a:152)

Against the sanctification she diagnoses in the vision of "every other" as "wholly other" in Levinas and then in Derrida,

Malabou offers change, mutation, and metamorphosis as primary and immanent. Indeed, if change is always outsourced to some absolutely other scene, how is it ever to be thought of as possible? In Malabou's view, change is possible precisely because there is no outside of transcendent alterity; this is the type of transformation that, "contrary to the claims of the philosophies of transcendence, can take place only in a closed system deprived of any relation to an outside conceived of as absolute alterity" (2012a:265–66; my translation). For Malabou, the elsewhere can only ever be right here and right now. No escape, nowhere to run—and for this very reason, the possibility of transformation.

This is what Malabou calls the "originary ontological fashionability of things" (2011a:90): "To come to be is to change. . . . [*M*]*utation nowhere begins*" (134–35). Being, she writes, "is perhaps nothing . . . but its mutability" (11), "nothing but (its) transformability" (72–73). If we can unhook ourselves from exteriority, from alterity as transcendence, we meet universal transformability and exchangeability. "If we accept . . . that alterity can be thought without the aid of transcendence, if it is true that there is nothing outside, nothing beyond, . . . then *there is no inconvertibility. Absolute convertibility, the migratory and metamorphic resource of alterity, is the rule. Absolute exchangeability is the structure*" (2010:46–47).

Programmatically, Malabou states, "*Metamorphosis—is what is. Migration—is what is*" (2011a: 199). But she formulates these tautologies precisely to undo the trope's usual implication: between "is" and "is," at the heart of the tightest formula of supposed identity, she finds mobility. Even the logical proposition that A = A presents identity only as the result of an exchange, the mediation of repetition as operated by the equals sign or, indeed, the verb "to be"—which accordingly for Malabou (here no doubt at her most faithfully, speculatively Hegelian) changes something into what it is and is not. There is no substance, nothing other than change—and nothing before or beyond change as its hidden motor. As Ian James writes, "Originary exchangeability, change and transformation are *as such* only in and through this primordial absence or void of essence, substance or ground" (2019:197).

There is for Malabou no unchanging essence, no metaphysical substance that might remain intact regardless of its accidental actualizations. Malabou calls this "substantialist assumption" "the travelling companion of Western metaphysics": "Form transforms; substance remains" (2015b:7; translation modified). Rejecting it, she is rejecting the metaphysics that would regulate change in advance by reducing it to the mere presentation of an invisible principle lurking offstage—in order to affirm every and any change as always possible. For this reason, if any one plastic form is not infinitely mutable (as we saw earlier), the possibilities of change are nonetheless limitless—and the resulting ontology is joyfully plural. "The new metamorphosis and migration results from the way in which all things, from here out, breathe the same air, and live the same novel ontological condition. . . . In the midst, again, of this gift form mobile, nomad assemblies, groups of resemblances and 'resemblants,' and clusters of metamorphoses" (2011a:174).

At the origin, there is (ex)change, and so also nonequivalence, including within (ex)change *itself*. (The mobility of tautology.) For Malabou, nothing equals anything else—because nothing is ever even equal to itself. Originary (ex)changeability *undoes* the indifference of general equivalence (the universal fungibility of everything—including ever more intimate aspects of all forms of life—into the money form, money as universal solvent): this is the meaning of Malabou's insistence on "sameness." "Nonetheless," she emphasizes, "*everything does not come back to the same*"

(2011a:236). Refusing to outsource disruptive displacement to some mythical "other," she configures a plastic commonality, "incipient forms, alien traffic, supple growths, and unmanaged exchanges" (161). Against general equivalence, then, Malabou offers "cineplastic *equivalences*: every form and displacement are of value" (161). Against general equivalence, and against a metaphysics of replication: "Modification," she insists, "is not the alteration of a preexistent form" (240). On the contrary: "Form can cross the line such that ontological transformability cannot be conflated with the possibility of fashioning or creating from scratch an essence, a god, a man, a thinking, a mode of speech, or a relation to being. This transformability is, rather, the revelation of an essential fragility . . .—the supple, changing core in each thing that does not allow itself to be seized hold of but that nonetheless is neither unattainable nor transcendent" (273–74). When form does cross the line—and it does—it thus becomes "the other of the idea—a non-ideal form that is at once both the condition and the result of change" (279). Which therefore offers always more possibilities, *without return*: no way back, a change of ground giving exchange instead of accumulation. In place of a return, the explosion of tautology.

DIS-CONTINUITY: THE ACCIDENT

Malabou's change, then, requires no transcendence. What is more, its drastic, at times explosive mutations do not themselves give rise to any transcendence; the breaks they form definitively reconfigure their milieu while remaining wholly immanent to it. If this refusal of transcendence aligns her to some extent with Latour, then, any such alignment can be only very limited; like Stiegler, she is distinguished fundamentally from Latour's position by her insistence on refusing transcendence while remaining committed to drastic interruption. In *Plasticity at the Dusk of Writing*, Malabou calls this "the simultaneity of *suture* and *rupture*" (2010:35) and, more axiomatically, "the dual impossibility of a *strict continuity* and a *clean break*" (29; translation modified). Though change itself disqualifies any "strict continuity," its determination as immanent means that it opens onto the mobile possibilities already swarming in the here and now, not onto some fantasy other scene—hence, no "clean break." It is, in fact, Malabou's contention that the understanding of rupture and continuity as incompossible itself prevents a rigorous account of immanent transformation, which is, in fact, characterized by their conjoint operation. Against such an understanding, Malabou insists on change as a dynamic of articulation, in which, as she writes in *The Heidegger Change*, "the point of rupture is also the point of suture" (2011a:283). Or, in a more dramatic moment from the same work, "the continuity and contiguity of constantly reformed forms as well as the emergency leap through the ring of fire" (125). In the terms we developed in discussing Stiegler earlier, this is change as immanent inflection and drastic interruption together—a watershed, a fold with the force of a cut.

In a sense, this understanding of change is already given in the concept of transformation, as long as we free transformation from any notion of the realization of a transcendent model (or, indeed, an ideal form). Thought in this way, transformation can happen only "right at the level of the form," as Malabou puts it in *The Future of Hegel* (2005:192); that is, "form cannot be 'undone' without calling on *the support of form*" (2010:46). That change comes from nowhere—from nowhere else, nowhere other than right here—means we have to think change via the logic of mutation and its process as an articulated dynamic of dis-continuity. As mentioned, Malabou's key motif for this way of understanding change is the accident: if she entitles one of her earlier works *Ontology of the Accident*, this is not, as it were, by chance. The importance of this figure for the overall model we are developing across this study, of course, makes her use of it especially interesting for our purposes. So let us now explore just how she does use it.

We have seen *The Future of Hegel's* interest in "the process by which a contingent event, or accident, touches on the heart of the system, and, in the same breath, changes itself into one of the system's essential elements" (2005:193). More than an interest, this is, in fact, a governing concern of the work: it opens the principle of plasticity Malabou generalizes from her reading of Hegel's *Phenomenology*. When we read a statement such as the following, toward the end of the text, then, we find Malabou both summarizing that reading and stating this principle: "Everything begins in the same moment, where the becoming essential of the accident and the becoming accidental of essence mutually imply one another. There is nothing beforehand" (164).

The two tendencies Malabou here describes as co-implicated—"the becoming essential of the accident and the becoming accidental of essence"—respectively characterize the first two moments of what Malabou presents as a (perhaps unsurprisingly) three-step movement in Hegel's account of the self-determination of subjectivity through the self-differentiation of substance, in which each moment gives form to its own distinctive temporality. The first of these is the teleological circularity of ancient Greek time, in which substance becomes subject through the repetition and internalization of habit as the constituting expression of individuality: the accidental becomes essential. The second is the new, linear temporality opened by Christianity as revealed religion, in which the movement of kenosis, the self-emptying through which God becomes Man in a singular event, sees subject become substance: the essential becomes accidental. (After these two, but less significant for our current purposes, the third moment is, of course, the properly philosophical moment of absolute knowledge, in which subjectivity determines itself through its knowledge of these two previous moments, whose temporalities might now be grasped as such, gathered and sublated in a new time in which time is neither circular nor linear but plastic.) In Malabou's elegant summary, "The Greek concept of 'form-creation' (*Formtätigkeit*) implies that self-determination is to be thought in terms of the becoming essential of the accident. . . . On the other hand, the modern concept of plasticity seems to align itself with a concept of self-determination as the becoming accidental of essence, the form of becoming which constitutes the fundamental meaning of the Incarnation" (119; translation modified).

If the core claim of *The Future of Hegel* is that Hegel's speculative completion of the time of history in the realization of absolute knowledge does not cancel all futurity but, rather, opens the possibility of immanent, plastic change, Malabou finds the form of this possibility in the double dialectic of these first two moments, their originary "co-implication" of essence and accident. Indeed, her understanding of plasticity as originary derives precisely from this co-implication: that there is nothing before or beyond change means that there is nothing outside of contingent occurrences and the continuity these both form and interrupt. This is why Malabou's summaries of Hegel toward the end of her study stress the first two moments over the third; describing Hegel, she is also describing herself. (And in so doing, she is, incidentally, performing what her study describes as "plastic reading": we will come to this later.) At this stage of her work, then, we might say that not only Hegel's but also—and perhaps especially—Malabou's philosophy "assumes as an absolute fact the emergence of the random in the heart of necessity and the fact that the random becomes necessary" (163; translation modified).

So far, so quasi-causal, perhaps: an ongoing, looping narrative of bootstrapping becoming through breaks. With *The New Wounded*, however, the emphasis shifts: away from the pole of continuity toward that of discontinuity. We have already seen that Malabou considers the destructive plasticity she identifies as at work in catastrophic cerebral lesions as effecting a radical break; we are now in a position to explore the precise nature of the discontinuity it introduces. If her discussions in *The New Wounded* still turn on the relation between contingent accidents and the continuity into which they irrupt, Malabou's increased attention here to specifically destructive plasticity means that she is less concerned than in *The Future of Hegel* with the dialectical equilibrium of the accidental rupture and its essential incorporation, than she is with the extreme pressure placed on the very possibility of incorporation by the severity of the accidents she is now addressing. As a consequence, *essence* falls away as a term, and *accident* becomes central in its own right, even defining the entire field Malabou is here exploring. ("Cerebrality," she writes: "the evental regime of the accident" [2012b:63].)

This also brings about a significant shift in the relation between the accident and its meaning. In *The Future of Hegel*, the accident cannot but become meaningful, inasmuch as its dialectical interplay with the essential constitutes the story of the subject and the story of philosophy itself. In *The New Wounded*, by contrast, the accident appears as a wild interruption of the order of meaning, which sits within the narrative of the subject's life as an inassmilable break without content. Such accidents are, Malabou writes, "hermeneutically 'irrecoverable' " (2012b:5). The psyche survives but is now radically interrupted: "*The cerebral accident thus reveals the ability of the subject to survive the senselessness of its own accidents*" (5). Inassimilable, the accident is now also irredeemable and challenges the very notion of survival: "Despite its name, such survival does not appear as sublation or redemption. It has nothing to do with salvation or resurrection" (60). The raw contingency of the accident, now without any dialectical relation to the subject's story, inaugurates a future unthinkable as the continuation or fulfilment of any past.

Malabou's continued use of the lexicon of survival as a description of the subject's relation to such catastrophic discontinuity is nonetheless a useful reminder that in her understanding, this wildly contingent rupture remains immanent within a form of continuity, producing the articulated structure of dis-continuity. There is still for Malabou no "clean break": even this severe disruption remains an event within a closed system, which neither arrives from nor leads to any transcendent outside. This is most distressingly illustrated by the cases of severe cerebral lesions discussed in *The New Wounded*, in which the person affected is now unrecognizable as the person they once were. As Malabou puts it in *Ontology of the Accident*, "a new, unprecedented persona comes to live with the former person, and eventually takes up all the room" (2015b:1). For Malabou, the pain of this scenario derives from its terrible refusal to hold discontinuity apart from continuity: the person is unrecognizable to the precise extent that they are also the same person, separated from their former self by a distance that remains internal. With reference to the paradigmatic case study of Phineas Gage, a late-nineteenth-century construction worker who suffered catastrophic brain damage in an industrial accident, Malabou cites Antonio Damasio's formulation of this paradox: "Gage was no longer Gage" (2012b:16). As Christopher Watkin writes, "in order for the very transformation and misrecognition on which Malabou insists to obtain, there needs to be a continuity of some sort from the person before the transformation to the person after" (2017:129). Watkin is in fact criticizing Malabou here, arguing that her focus on the neurological constitution of subjectivity neglects the social embeddedness that here permits a sense of continuity: if Gage is also still Gage (as suggested by the persistence of the name), it is because he is still in some sense recognized as Gage by those around him.

Of course, the embedding of the interruptive accident in a medium of continuity is already part of Malabou's overall model of transformation, and, indeed, it is indicated in this case by her image of two cohabiting personas. And if the interrupted continuity this displays appears paradoxical, that can be so only in relation to our habitual insistence that interruption must crash in from elsewhere. Malabou challenges us to rethink this; in her terms, the "new" Gage is no less radically other for having emerged not as the manifestation of some transcendent template but from within the neurology of the "old" Gage. In a sense, Malabou is only asking us to be rigorous. If we want to give up a metaphysics of ontological transcendence (and so any dream of an "elsewhere"), we can either accept that this also means giving up on drastic discontinuity (if we continue to believe that radical interruption can come only from such an elsewhere), or we can agree that such discontinuity cannot be other than processually immanent. This is why her conception of dis-continuity—that is, of destructive plasticity as drastic discontinuity within continuity—is more than a paradox: it is, rather, a confrontation with the implications of refusing ontological transcendence for any consistent conception of transformation.

If we now bring *The Future of Hegel* and *The New Wounded* together, we get Malabou's delineation of the accident as compacted continuity and discontinuity within the temporality of such transformation. The tendency to continuity pulls toward the pole of complete narrative integration; the tendency to discontinuity, toward the pole of pure evental puncture. And the accident? The accident marks the place of any transformation whatsoever as always taking place somewhere along the spectrum between these two absolutes. We are now in a position to address Malabou's full importance for the model of decisive, distributed political agency that I am developing in this study. For it is in terms of such dis-continuous transformation that Malabou thinks the emergence of an agent as a site of responsibility and futural commitment. The emergence, that is, of the accidental agent.

PLASTIC READING

We find this account of accidental agential emergence in the key section with which Malabou concludes *The Future of Hegel*, which is dedicated to the appearance of the figure she calls the "plastic reader." As we now explore this figure, we will remain within Malabou's framework, in which the agency in question is posed in terms of hermeneutics—of reading and interpretation. Malabou's publications immediately after *The Future of Hegel* make this framing surprising, in retrospect: more than once in these texts, she presents the paradigm of plasticity as taking over from hermeneutics, as the appearance of "a new motor scheme, the accomplice of a new era" at "the dusk of writing" (Malabou 2010: 17). And, indeed, as we have seen, the concept of destructive plasticity in particular insists that the transformation in question explodes any possible recuperation into schemas of meaning-making.

Why, then, tarry with what might be just a residue of Malabou's dialectical and deconstructive background, a last farewell to philosophies of interpretation? Simply: because the overlap in the figure of the plastic reader between the accidental emergence of plasticity and the interpretative venture of reading will be key to Malabou's contribution to the development of our model. If the historical status of plasticity in relation to hermeneutics need not concern us greatly here, it will be of considerable significance that Malabou's plastic agent *emerges* as the locus of an *interpretation*. For, as we will see, it is this that allows her to describe this agent as engaged in a commitment to the future. And this conjunction of emergence and futural commitment is at the heart of Malabou's importance for our purposes.

From the outset, Malabou defines plastic reading—the experience of reading offered to Hegel's reader after the apparent completion of the self-realization of spirit in the moment of speculative philosophy—as the intersection of freedom, responsibility, and the decision. "The philosophical decision finds itself in a new modality. . . . This is the freedom and responsibility of *interpretation*" (2005:168). Faithful to Hegel's definition of speculative philosophy, Malabou emphasizes that these aspects emerge through a process: here, the process of reading, specifically the rereading that, for Hegel, is required by the speculative proposition. Most important here for Malabou is the looping movement by which the reader takes responsibility for a meaning they have not created by assuming their position as the site from which a new proposition must be produced. In Malabou's account, this process first entails the disappearance of the reader as a "particular" self—from which gulf they emerge in the act of assuming both this disappearance and their responsibility for the meaning of what they are reading. As Malabou describes it, "In the first stage, the reader does indeed let go of every and all form: . . . the reader loses the form of the 'knowing I', which had represented the self. . . . And with this, the reader, as a 'particular

Self', disappears. . . . In the second stage, the reader must recover from (*se relever de*) this decline so that the effective arrival of the proposition, the assumption of its meaning, can show itself" (180; translation modified).

Reemerging in this way as the site of a responsibility for a meaning of which they are not the origin, the reader appears in the act of reformulating this meaning in their response: "Far from dissolving into the content of what is read, the reader must in return express this content and, to do so, can only formulate new propositions, transform the content of the reading, interpret it" (180; translation modified).

Crucial in this respect is the hiatus, the interruptive non-place of the accident that, in this process, comes retroactively to occupy the place of the origin. Of the Hegelian reader, Malabou writes, "But at this place of return, the reader finds *nothing*. As the origin was never there a first time, the reader cannot discover any substantial presence, any substratum waiting to be identified" (2005:179; translation modified). The reader comes into existence accidentally, the location of a transformative interpretation occasioned but not determined by what is read; in the place of the origin, there is nothing but this impersonal emergence. "There is nothing beforehand" (164). No substance to program form, but dis-continuous transformation. As we are here interpreting Malabou's account of plastic reading as a model of the formation of agency, however, this emphasis on accidentally emergent singularity as the dissolution of particularity—and, indeed, individuality—gives rise to a potentially awkward question. As she puts it, "we might find ourselves objecting: what kind of status can be enjoyed by this interpreting subjectivity if, as our analysis has shown, the Ego has been dismissed? Doesn't the 'Self,' because of its anonymity and its automatism, threaten to annihilate all individual exegetical initiative?" (2005:168).

Given that the interpretation offered up by the reader is no longer simply "theirs," in what sense can we talk here of agency, let alone the responsibility it would serve to guarantee? Malabou's response is to clarify the conception of responsible agency that she is proposing. As she formulates it, "it is indeed to the individual reader that it falls to present, in return, the movement which, paradoxically enough, led to the collapse of the 'knowing I,' hence of individuality itself" (2005:81; translation modified). For this is only a problem, in fact, so long as we insist on attributing meaningful responsibility solely to an agent in whom the capacity for such responsibility would be inherent. This metaphysical model of a fixed subject as the support of given predicates is, of course, what Malabou's commitment to plasticity emphatically rejects—and it is what she designates here under the name of "individuality," as the *particular* instantiation of a transcendent form. Against this, her task is accordingly to present the *singular* agent that emerges from the process of plastic reading as the locus of meaningful responsibility, precisely inasmuch as it is now produced in the moment of its exercise, not enacted as an already determined capacity. (Here, Malabou doubtless recalls Derrida's account of the would-be just decision, in which to seek to be just, the decision cannot merely apply a preexisting code; see Derrida 1990.) Malabou writes, "The letting-go of the Self in the act of reading does not suspend all power of decision; on the contrary, it produces its condition of possibility" (2005:182; translation modified). The emergent reader-agent is in no sense excused from the need to decide on the interpretation to be proposed. On the contrary, it is just because the formulation of this interpretation is the act in which this agent emerges as responsible for the meaning it transforms that this formulation constitutes a decision.

There is, then, no essential or necessary relation between the emergent reader-agent and the meaning through whose interpretative exposition and transformation this figure emerges; as the absence of any preexisting ground implies, the relation is purely accidental. In Malabou's terms, "Interpretation is a *production* that takes responsibility for the accident which gave it birth" (2005:185; translation modified). Moreover, sending its propositions out into circulation in their turn, as a newly transformed meaning, this production is also turned toward the future. Taking responsibility for its emergence in an act of interpretation, the reader-agent accepts that this interpretation "cannot be definitive, but is itself promised to other readings" (185; translation modified). Its future is that described by Malabou in the motif she calls *voir venir*, neither utterly opaque nor wholly transparent, a dimension of immanent exposure and commitment. "The future is not the absolutely invisible, a subject of pure transcendence refusing any anticipation at all," writes Malabou. "Nor is the future the absolutely visible, an object clearly and absolutely foreseen" (184; translation modified). If the future to which the plastic reader commits its interpretations thus imposes, as Ian James writes, an uncertainty that undoes any transcendent mastery (2012: 88), it also declines any transcendence itself—the reader-agent is immanently oriented toward it, in the flow that will, in turn, bring unpredictable new accidents.

The accidental emergence of the reader-agent is thus poised between the two extremes of Malabou's plastic spectrum, neither of which can quite claim it: at one end, seamless continuity (full retrospective appropriation by a narrative of identity formation); at the other, pure discontinuity, the catastrophic destruction of annihilation. Both would negate accidental emergence—but the accident is precisely plastic change as irrecuperable by either pole.

That there is no seamless continuity means the accident cannot be integrated into a master narrative, absorbed and digested as a welcome taste of the exotic outside; hence the annulment here of the "individual," "personal Self." That there is no annihilation means, conversely, that the situation within which the accident arises continues, its ongoing alignments and allegiances defining the stakes of the meaning to which the plastic reader-agent will commit in the very act of its future-oriented emergence.

This plastic reader-agent is wholly defined as the position at which an interpretation is articulated and offered for future circulation, this offering constituting a commitment to take responsibility for this interpretation. There is no preexisting substance, no ideal form, no lurking subject that would somehow constitute the ground of this composite act. "There is nothing beforehand"—always only the void at the origin, abyssal groundlessness as the site of emergence. In Malabou's own account, her work as a whole seeks to situate a resistance to the "individual subject," a resistance located in the place of "subjectivity" itself: in the "indifference," "void," or "blank square" that constitutes "the 'essence' of subjectivity, that is its 'reality' " (2018: n.p.). If, for Nisbet, Malabou thus "importantly unhinge[s] the subject from human consciousness and radically open[s] the range of possibility for thinking change in a world" (2018: n.p.), I would therefore go further and claim that what she offers here, in fact, is the possibility of unhinging committed change not only from human consciousness but also from any language—such as that of subjectivity—that programs change as the implementation of a script from elsewhere. Freed from such substance-metaphysical determination, Malabou's model of plastic reading offers the possibility of dis-continuous transformation, as accident and decision, emergence and responsibility come together in a moment of processually immanent futural commitment.

"CLUSTERS OF METAMORPHOSES"

Granted, Malabou's model might appear curiously limited for our purposes. Restricting itself to reading, and the reading of speculative philosophy at that, does it not reinstall precisely such substance-metaphysical—indeed, anthropological—determination? If it is difficult to avoid this reservation, however, we need not give it much weight here. In the first place, of course, we might recall that with the cybernetic redefinition of "meaning" as "information," and its take-up in the biological sciences, "reading" ceases to be an exclusively human activity. But those who want to safeguard a strong version of reading, interpretation, and the commitment to future circulation that these enjoin as indeed exclusively human can always dismiss it as merely a more or less metaphorical extension of the term's semantics. We will spend more time with this question in the conclusion. For now, though, I want to emphasize that what matters for our purposes is not the anthropological framing that might or might not appear to be mandated by Malabou's model. What matters is the structure of this model—precisely because, as I will now argue, this structure is indifferent to such framing. As we have seen, Malabou's elaboration of a model of agential emergence via the image of plastic reading is not beholden to the metaphysical commitments this term ought to bring with it—which in turn allows us, in our own moment of plastic interpretative transformation, to export this model beyond the limits to which this image might have seemed bound to restrict it.

The account of agential formation I want to derive from Malabou hangs on the compatibility between her thoroughgoing commitment to immanence and a meaningful conception of political agency. For various of her critics, however, the former simply blocks the latter, leaving Malabou's invocations of freedom through plasticity disappointingly empty. With reference to *What Should We Do With Our Brain?*, for example, Katherine Hayles wonders how Malabou's modeling can "bring about the agential freedom she envisions," asking bluntly, "what practical actions are possible that would take account of this freedom?" (2012:82–83). And Alexander Hope states simply, "Malabou does not really demonstrate how this emancipation might take place" (2014:339). For Hayles, in fact, Malabou can rescue the effectivity she wants for her model only by appealing for increased consciousness of the possibilities offered by neuroplasticity. Ironically, this would lead her to reinstall precisely the split between the cognitive-symbolic and the neuronal-material she wants to refuse (2012:101–2). (For extensive further coverage of these arguments around the politics of neuroplasticity, see especially Gratton 2014; Moten 2015; Proctor 2011; Sparrow 2014; and Watkin 2017.) More generally, Malabou's refusal of transcendence is often seen as confining her ontology to deathly stasis: as Peter Gratton writes, for example, her "totalizing" account of immanence "suggests a one-All with absolute monadic interiority, where there is no Other to the system" (2014: n.p.). In Malabou's view, however, her position offers a resolutely materialist account of change—and, indeed, only such an account can maintain change as a possibility.

Malabou's inspiration for this position is the co-immanence of the biological-material and the symbolic that she finds in the neurobiological understanding of meaning-making, precisely because the brain is her paradigmatic mobile closed system, as we have seen. This is the empirical motivation for her rejection of transcendence: the freedom inscribed in the unpredictable, contingent elaboration of meaning is already immanent to matter, in an "originary intrication of the biological and the symbolic that never requires a transgression of the biological itself"

(2015d:44). The indeterminacy that the transcendence of the material-biological by the symbolic had sought to preserve in the name of agential freedom can now be understood in a way that renders this split—and the consequent instrumentalization of matter, and of all those beings defined as "mere" matter, devoid of or deficient in "spirit"—null and void. Instead, this indeterminacy can now be recognized as "the transformative tendency internal to materiality, the self-transformative tendency of life," as "life transforming itself without separating itself from itself" (2015d:45). Feared by those who secure freedom by reason's transcendent sovereignty, for Malabou this, on the contrary, inscribes freedom within the materiality of all forms, including those of thought, by virtue of the irreducible variation of their local becomings. Refusing both the outdated determinism of the genetically programmed phenotype and the unchanging abstractions of the transcendental, Malabou situates freedom in the immanent mutations shared by the cognitive-symbolic and the material-biological. Citing Foucault, she argues that " 'what is most material and most vital in bodies' must be thought as an interactive space, a formative and transformative dynamic of organic identity that operates *within* the economy of the living being itself, not outside of it" (2016c:433; quoting Foucault 1978:152). In the phrase that becomes something of a slogan for her position, "There is but one life, one life only" (Malabou 2016c:438; and for a critical response, see MacLeod 2016.)

Malabou is clear, moreover, that in understanding freedom in this way, she is also thinking about specifically political agency. She makes this particularly clear in the preface to the second French edition of *What Should We Do With Our Brain?* Again invoking Foucault, she argues that conceiving of resistance in terms of a dialectical opposition guaranteed by a transcendent other space (the truth of history, the deliverance to come, this or that essentially human quality as escaping the law-bound realm of necessity) will simply be absorbed by the dominant order, and that as a consequence, "resistance must take the form of an internal strategy, a dynamic that is immanent to relations of force, . . . from which it is impossible to escape or to free oneself. Any supposed escape from these relations will be assimilated into their movement" (2011b:24; my translation).

To insist on the possibility of a radical outside is thus for Malabou to remain attached to "a sort of transcendence fetish" (2011b:34; my translation). As the current dominant order, global capitalism is doubly happy to ingest such fantasies, claims Malabou; first, because they in fact reconcile its would-be subjects to the present, sustaining them via a dream of happiness to come, second, because the projection and colonization of virgin territory remain fundamental to its constitutive dynamic. As she summarizes, the search for a political alternative elsewhere expresses the "will to find, beyond what is, a new symbolic promise—which will never be anything more than a new illusion, which capitalism will make short work of" (32; my translation). The key term here is "symbolic": for Malabou, liberation will not be found through systems of meaning that ground their values in their transcendence of material immanence. Alternative reworkings of the symbolic can only feed a capitalism founded in the distinction between meaning (value) and matter (raw materials, fixed capital, labor power). "So long as we think that the answer to material immanence is to be found in the symbolic, then what we are opposing to capitalism is a mind-set—which will never ultimately be anything other than the mind-set of capitalism itself" (32–33; my translation). The separation of meaning from matter is the very mechanism of surplus value and the topsy-turvy logic of the commodity: this separation will hardly save us from an economy it has so faithfully served.

In the context of global neoliberal capitalism, however, Malabou's insistence on thinking political agency within immanence—particularly her refusal to index change to transcendence in either space or time—can look like anything but a commitment to liberation. As we saw earlier, her affirmation of "the form of a world without any exteriority" (2010:67) can look worryingly like the endlessly renewed (and endlessly self-defeating) fantasy of the "end of history." And, to be sure, Malabou does seem to adopt this thesis. In *The Heidegger Change*, for example, we find her asking, "What is change—how to change—when history is no more?" (2011a:279) Presenting the disappearance of "historical being" (282) as a universally acknowledged fact, Malabou appears to approach the triumphalist version of the "end of history" thesis most infamously formulated by Francis Fukuyama in 1989 (and dismantled by Derrida in *Specters of Marx*). In Malabou's case, however, this apparent alignment with the notion forms part of a commitment not to stasis or quietism but to change. She sets out this position at length late in *The Future of Hegel*, where, in a sense, it comes as no surprise; her argument in this work is that although the identification of speculative idealism with the world-historical seems programmed to reproduce the schema of post-historical exhaustion (which already imagined itself to be Hegelian, of course, derived as it was from Kojève's various iterations of the thesis in his commentary on Hegel's *Phenomenology*), this identification in fact offers the irreducibility of plastic change as the properly Hegelian future. Accordingly, when Malabou explicitly connects this scenario to contemporary geopolitics, this dynamic of futurity in exhaustion is precisely what appears.

With the globalization of capitalism, claims Malabou (perhaps with a nod to Kant's characterization of space and time as the pure forms of sensible intuition that enable any human experience of the world), time and space are experienced as simultaneously saturated and empty. The times of repetition and of the future as promise are over; and there are no more spaces waiting to be colonized. There is, supposedly, nothing left to do. In this mythology of completion, Kojève's cool ironies meet Carl Schmitt's disgust: "How terrifying is a world where there is no foreign land but only inner territory; no path toward the open; no leeway where forces are freely measured and experienced," wrote Schmitt in 1947. By 1978, he could define the "appropriation of world space" as (in Peter Szendy's gloss) "the end of politics, . . . its dissolution into a world police force in the service of the economy" (Szendy 2013:25–26; respectively citing Schmitt's *Glossarium* and his "*Die legale Weltrevolution*"). As Malabou puts it, "The 'new world order' means the impossibility of any exotic, isolated or geopolitically marginal event. Paradoxically, this saturation of theoretical and natural space is felt as a vacuum" (2005:192).

Against this supposed impossibility, and so also against those who, like Schmitt, can define politics only in terms of the transcendence and appropriation of finite space, Malabou's founding insistence on plastic change as irreducible and immanent allows her to find liberating potential precisely in the "saturation" and "vacancy" of "post-history." For it is, of course, definitional for Malabou that "plasticity designates the future understood as future within closure" (2005:192). "In this way," as she puts it in *Plasticity at the Dusk of Writing*, "existence would be the plastic drive within the closed form of the world, within the closure of globalization" (2010:71). Rather than refuse the notion of global saturation, Malabou acknowledges its force and asks us to think transformation from here out. It is all too easy to miss the combative nature of this stance; Malabou's emphasis on change as always possible is here explicitly an affirmation made *in and against* definitive closure. Plastic change as immanent strong discontinuity means accepting that change is not going to come from somewhere else—that transformation can happen only from within an existing form. From within the form of this world—whose alleged closure can, as a consequence, only ever be a myth and open to change.

Malabou's refusal to define freedom as the escape from or overcoming of immanence takes us to the heart of the questions generated by a politics without transcendence. For her position calls into question the very division in which the modern conception of the political subject has its abyssal foundation: the "Great Divide," as Latour calls it, between the natural and the cultural, necessity and freedom, the material and the symbolic. That is, the separation of the political from everything it has been thought necessary to exclude from this realm in order that it might establish itself, a separation that itself at once guarantees and is guaranteed by the exceptional status of the human being as transcendent in relation to this *everything else.*

In the criticisms of Malabou's philosophy for its supposed political inertia, we can see something of what is at stake in attempting to think politics without this ontological transcendence—and so, what is at stake in this study as a whole. If I hope to have shown so far that it is indeed possible to think political action in processual and emergent terms as the decisive distributed agency of ontologically diverse alliances, these criticisms suggest how tenaciously our dominant understandings of politics cling to the freedoms supposedly secured for human actors by their infinite distance from all other kinds of being. For Malabou, by contrast, if the freedom thus gained for this political actor is grounded in the supposedly inherent properties of the kind of being called human, then it is in fact just a metaphysical version of the determinism it imagines itself as transcending. In refusing to ground freedom in the separation of the symbolic from the biological, as the transcendent attribute of a given type of being (and so a gift from some munificent beyond), she is insisting on freedom as emerging accidentally in its performance at this or that site within an evolving symbolic-material system. Without such accidental emergence, we have, at best, fantasies of transcendence backed up by the stasis of the transcendental. With it we have the freedom to be in process, which is to say, to change.

Neither the stooge of transcendence nor deathly confinement, Malabou's plastic immanence is already a mobile, plural realm of possibility and differentiation. What is more, neither we nor she need prise this realm open with the crowbar of exceptional human subjectivity in order to understand it as also the realm of meaningful agency. Malabou already has a fully functioning model of such agency, at once processually immanent and responsibly future-oriented. Sparrow is quite right in his assessment that in Malabou, "The dialectic of formation and explosion is meant . . . to salvage a conception of agency" (2014:227); and this conception operates regardless of any appeal to substance-metaphysical grounding. Like Stiegler, Malabou is, in fact, much closer to process philosophies than their shared post-Hegelian inheritance might suggest. If we recall Stiegler's use of Simondon, for example, we might say that for Malabou, as for Stiegler, "the agent" names a form that—like all forms—crystallizes and deliquesces in dynamics of individuation and disindividuation; and that if the form that goes by this name is identified in her work as a locus of future-oriented commitment and responsibility, it does so while remaining wholly immanent to the process of its emergence. It is this immanence that allows the agency modeled in plastic reading to escape the pull of anthropological determination. Malabou's model describes nothing more and nothing less than the accidental emergence of a site of interpretation that engages a commitment to the future by putting forth this interpretation in the very act of its emergence.

We began this chapter needing an account of agential emergence that might be able to accommodate the quasi-phenomenal, semi-opaque distributed political decisions of digital-human agential alliances, while also accounting in detail for the futural dimension of partisan commitment introduced into our model by Stiegler. I want to claim that Malabou's plastic reading has given us such an account. Distributed digital-human decision-making gives its participants no access to the whole of the process in which they take part; they emerge as agential sites in the act of putting forth an interpretation as their contribution to a composite intervention for which they bear a share of responsibility without definitive authorship. Not authors, but plastic readers. No agency without the empty, abyssal nonorigin; no foundation in retrospective appropriation, because there is nothing there to appropriate. What is assumed, then, in agential emergence? Precisely this: that the agential alliance emerges together with its participants in a process itself unavailable for appropriation, as the plural location of a composite decision ungrounded in transparent self-consciousness, a futural commitment that takes responsibility for its partisan alignment within a conflict over value without grounding this responsibility in supposed a priori qualities. This is why Malabou's conception of emergence and responsibility—like Stiegler's—is more radical than the existentialism its emphasis on a freedom composed of ungrounded, heteronomous, situated decisions can seem to recall. Here we have responsibility as also free from determination as *merely* human. As action is composite, so, too, is responsibility: resting with the alliance as a whole and also parsed out differentially, according to local contributions and capacities, themselves emergent thanks to their relative position within the alliance and far from mutually transparent. Quasi-phenomenal responsibility. “Responsibility for our natural irresponsibility,” as Malabou calls it (2015a). Differentially distributed across an alliance, this is responsibility as eminent accountability for the act of participation in which the agent emerges thanks to the contributions of its allies; and as a commitment to the meaning of this act, which crystallizes, inflects, and recirculates the stakes of its milieu.

Malabou’s contribution to our model of distributed decisive political agency is thus twofold. In the first place, her understanding of plastic change, especially in its explosive form, has provided a fully developed schema of immanent strong discontinuity, through which we can think decisive intervention as transformative interruption without needing to index this possibility to some kind of transcendence. Stiegler gave us the bifurcating decision as immanent recursivity, reflexivity without transcendence, and, indeed, processual discontinuity; Malabou has given us a detailed theorization of such drastic change as inflection and interruption together. And if, like Stiegler, Malabou thinks this dis-continuity as the site of agential emergence in an act of decision, her description of this act as an interpretation offered to future circulation has provided a detailed way to understand how this emergence also entails futural commitment and responsibility.

If we now reintroduce the transvaluative dimension of Stiegler’s quasi-causal decision, we can see this futural orientation as a specifically partisan commitment within a conflict over value: the agent emerges in an act of decision that interprets its milieu in this way rather than that way, in favor of this tendency and against that, as an intervention affirming these values against those. And if we now, in turn, reintroduce what we saw in horizons 1 and 2, we can recall, first, that this emergence is not that of an isolated, self-sufficient agent but, rather, that of an ally, transformed from a being in an array to a participant in an antagonistic alliance precisely by this act of normatively committed interpretation; and, second, that the composite agential front of this alliance is made up of nonhuman and human participants whose respective capacities and responsibilities emerge relative to their different positions within this participative action.

Malabou has, then, provided a further dimension to our model. But we will have to develop her contribution further if it is to prove wholly compatible with our endeavor here. For as the figure of the plastic reader indicates, Malabou's account of accidental emergence is invariably conjugated in the singular. If the singular names the empty space that emerges from the transformation of the particular and the individual in the dynamic of plastic reading, this emptiness does nothing to complicate this space's unitary configuration. To bring the development of our model to its completion, we will accordingly need to open this unitary space back up, insist again on its plural composition. For the accidentally emergent, plural agential location we are interested in can emerge at all only in the form of an alliance. Pluralizing Malabou's emergent reader-agent, we will need to multiply this figure until it resembles what she celebrates in *The Heidegger Change*: "mutant kin," "clusters of metamorphoses" (2011a:285, 174), transformations forming alliances from among arrays of disparate beings, as mobile composite sites of decision and commitment. To conclude the elaboration of our model, this is what we will now do.

CONCLUSION

In unpacking the key features of Malabou's plastic reader, I referred to this figure by the broadly neutral but strongly humanizing pronoun "they." But the plastic reader is not human. It is the location of an operation. Malabou writes of Hegel that what interests him in the human is not some supposed substance, "human-ness," but its "status as an insistent accident" (2005: 73). We could equally say that what is of interest in Hegel-Malabou's plastic reader is not a capacity indexed to this or that substantial determination of a given being but the operation it describes: namely, the emergence of a locus of interpretation as a decisive immanent orientation toward the future. Such nonhuman readers today have, in one sense, become familiar figures: once the cybernetic redefinition of reading as no more or less than the interpretation of information exploded the theater of intentionality, leveling the field it thereby expanded, and with the enthusiastic take-up of this redefined figure in theoretical biology, readers appeared everywhere, from ribosomes working on mRNA to biometric sensors or air pollution monitors. In the words Latour lends to those scientists who, around the turn of the twentieth century, came across the existence and agency of microbes: "There are more of us than we thought" (1988:35).

Not all readers are plastic readers, however. Not every act of decoding composes a commitment. The scanner that reads my card when I buy groceries does not, in so doing, adopt any particular stance toward the future (however significant its results might prove for my own). But we should not conclude from this that interpretation as futural commitment is the exclusive preserve of human beings. It seems churlish, for example, to refuse to think of trees connected by mycorrhizal networks as plastic readers. Over and above the mutual benefits of symbiotic relationships between trees and fungi, the "wood wide web" (Giovannetti et al. 2006; Macfarlane 2016) or "fungal superhighway" (Barto, Weidenhamer, Cipollini, and Rillig 2012) serves primarily to transfer nutrients and minerals (from areas of relative abundance to areas of scarcity) but also to communicate chemical warnings and other signals. It plainly thereby links sites of interpretation, which, through these interpretations, emerge as enacting a commitment to the future. When trees shaded by the leaf canopy take up the extra nutrients released by a taller neighbor, the latter has engaged itself in promoting their future growth. Or, in a more extreme instance of this scenario, when myco-heterotrophic plants, which are unable to photosynthesize, find the energy they need through such carbon transfer, the donor plants in question are favoring a future in which those others survive. I see no compelling reason not to understand these processes as instances of plastic reading.

Not all transfer is unequivocally benign, however; precisely as a commitment, such action might favor the development of seedlings from the parent species to such an extent that positive feedback loops result in monodominance on the part of the species in question (McGuire 2007). And when we focus on single beings (this tree, this retinal scanner), species (Douglas firs, infrared scanners), or, indeed, kingdoms (plants, electronic sensors) as *also* exhibiting this or that capacity, we promote our own habitual form of monodominance by the transfer of the substance-metaphysical attributes of human beings to beings of other kinds. (This is the irredentist humanism that continues to characterize various supposed posthumanisms, as we saw in the introduction.)

Rather than celebrating the admission of various other beings to the reading room of our exclusive club, then, we might learn the lesson of the wood wide web: plastic reading is the composite action of an alliance; in this case, the common mycorrhizal network in question. The emergence of a locus of interpretation in an act of futural commitment takes part in an immanent inflection of tendencies flowing through the network. With this inflection, elements within the array making up this network equally emerge as participants in an alliance whose contributions and vital interests shape the inflection as decisive collective endeavor. And when the decision enacted by this inflection adopts a committed stance toward the future, it indicates the partisan alignment of these participants along the alliance's agential front.

Is this language of decisive interventions and agential fronts exaggerated here, with reference to trees and fungi (or, indeed—recalling horizon 2—microprocessors and digital algorithms)? Is this all just so much pathos, not to say bathos? Perhaps. It is worth recalling, however, not only that the chemical warnings and spare resources channeled through mycorrhizal networks often constitute responses to situations of extreme peril but also that, as it multiplies such situations throughout all manner of lifeworlds, the generalized ecological emergency evoked under the contested name of the Anthropocene defines the vastly diverse forms in which these dangers are registered and opposed as contributions to a broad alliance, aligned in a confrontation whose stakes are no less than existential threat. In what meaningful sense are, say, mutualist networks of rainforest trees and fungi, seeking to promote each other's survival, *not* allied with the forest guardians of the Guarani-Kaiowá and the Xakriabá peoples and other activists in the fight against murderous deforestation? Humans versus Earthlings, in Latour's terms, already in actual, frontal conflict. Remember, "They, our adversaries, mobilized long ago" (Latour 2015: para. 20).

With this context in mind, I propose now to draw this study to a close by underscoring key elements of the model of distributed political agency that it has sought to develop. After a brief narrative summary of this development, I will conclude by synthesizing the normative dimensions of this model and suggesting just what it has to offer as a way of understanding politics.

Chapter 1 established the study's principal concerns: the interest of understanding agency with Latour as an emergent property of composite alliances made up of participants of all kinds, and the importance of expanding this understanding in the direction of greater political effectivity, adding possibilities of confrontational alignment, futural commitment, and prescriptive mobilization. Of demonstrating that distributed agency can also be decisive, in the two senses that have run throughout our explorations: making a major difference to a state of affairs and taking a position, engaging a partisan commitment.

Horizon 1 then proposed the figure of the antagonistic alliance, formulated to suggest both the compatibility of distributed agency with political conflict and the decisive quality of such composite action, in the first of these two senses. With that in place, chapter 2 began the work required to secure the second of these senses without appeal to human-exceptionalist fantasies of ontological transcendence, following Stiegler to conceptualize the decision as a partisan commitment within a conflict over value and as processually immanent recursivity, promoting a tendency in a moment of interruptive inflection as reflexivity without transcendence. Our model then began to engage substantively with the question of normativity and opened to the future, as it became possible to understand the projection necessary to such normative commitment independently of human-exceptionalist ontological transcendence.

In horizon 2, this account of the processually immanent decision rejoined the distributed political agency developed in horizon 1, as the quasi-phenomenal political decision-making of the digital-human alliance defined its sometimes starkly interventionist decisions as normative commitments remaining both irrecoverably distributed and wholly immanent to their situation. At this point, the remaining task was to understand just how such immanent, decisive, distributed political action could engage such a futural commitment. For this we turned to Catherine Malabou, whose conception of plasticity both allowed us to appreciate the possibility of decisive change in a situation devoid of transcendence and introduced the figure of the plastic reader as emerging in a moment of such immanent strong discontinuity as the site of a futural commitment to the value affirmed in their act of interpretation. And then, at the start of this conclusion, we pluralized the plastic reader: because it is the alliance that acts. Nothing and no one ever acts alone.

In solidarity especially with Latour and Bennett, then, I have argued that agency is an emergent property of ontologically diverse alliances. In the version of this model that I have elaborated here, participants in these alliances emerge as local positions along an agential front, with a share in the agency of the alliance, and exhibit differentially distributed capacities thanks to these respective positions. What I have particularly sought to add to the transformational contributions of Latour and Bennett, of course, is an emphatic and consistent characterization of such ontologically diverse alliances as antagonistically engaged, of that agential front as the line of a partisan confrontation, and of the alliance as enacting in this partisan alignment a composite futural commitment. In addition to its redescription of the scenario of agential emergence, then, this study has in particular proposed an account of decisive, distributed political agency.

I will come back to this emphasis on antagonism shortly, as we draw to a close by considering further what this model might offer to an understanding of politics. To reach this point, however, we will need to dwell a little longer on the question of agency and revisit the figures of the accident and the decision in order to sum up the normative dimension of our model.

Agency never was the "agency ex nihilo" we dreamed it might be (Rammert 2012:91), never followed the scripts we invented for the theater of intentionality. Look where we will, our fantasies of exclusive sovereignty remain unfulfilled, as the distance narrows between the decision and the accident. Transcendentally or empirically, we find our executive decisions looking increasingly like ratifications of initiatives launched somewhere upstream. Even for those who still prefer to cloister the whole of this drama within the limits of the human individual, decision neuroscience is parsing human decision-making out into ever more fractional nonconscious microprocesses. Empirically, pretty much any decision you care to name is starting to look more or less quasi-phenomenal, taken to some extent before, below, beyond, or outside the belated stagings of self-consciousness. Hayles quotes Dresp-Langley: "A great deal of human decision-making in everyday life occurs indeed without individuals being fully conscious of what is going on, or what they are actually doing and why. Also, human decisions and actions based on so-called intuition are quite often timely and pertinent and reflect the astonishing ability of the brain to exploit non-conscious representations for conscious action, effortlessly and effectively" (2017:56; citing Dresp-Langley 2012:7; and see Shadlen and Roskies 2012).

Even within the individual human being, then, we are perhaps starting to think of decision-making less as the sovereign authorship of a projected action and more as the emergent futural commitment of a plastic reader, as "getting behind something," "what a body gets behind" (Ahmed 2014:37, 142): as a partisan engagement within an ongoing process. As Bennett writes, "From a process-forward perspective, then, a decision is not the apex of judgment but a (more or less useful) blockage of its course" (2020:48). A watershed, inflection, and interruption together. The decision as heir to the accident, emerging from an abyssal nonorigin, surfaced from a flow it has no hope of synthesizing, from within which its emergence could not have been predicted but which it also reconfigures, tying itself in the act of its appearance to the value of *this* tendency. And for those who lean more toward some form of broadly collective, situational constructivism, somewhere on the socio-techno-biological spectrum, Bratton advises as follows, in similar vein: "What at first glance looks autonomous (self-governing, set apart, able to decide on its own) is, upon closer inspection, always also decided in advance by remote ancestral agents and relays, and is thus automated as well. . . . If moments of crucial choice depend on everything that is already routinized then 'the political' would refer not only to those rare self-conscious executive choices, but even more so to all those choreographies of technically entrenched pathways" (2019:39).

The key word here is *also*. Contrary to human-exceptionalist panic, the point is not that decisions are not taken in the soliloquies of the "I think" but that they are not taken here *alone*. Bratton's scenario also reminds us that the capacities suggested by these soliloquies are themselves already defined as such partly by their genealogy (the imbricated, historical coconstitution of the technical, the social, and the biological) and partly by their differential participation, alongside all manner of more and less sentient beings, within wider agential alliances. And it is the alliance that acts—with these self-conscious decisions as one of the contributions to its decisive, sometimes conflictual interventions. Accidental, yes—emergent from endless upstream regress—the alliance is also decisive; decisive, in fact, precisely thanks to this immersion, the thick, composite existential motivation that defines the intensity of its intervention and raises the stakes of the partisan commitment in which it emerges.

We will return shortly to the question Bratton raises of "the political." But first, we must complete our consideration of the implications of this way of understanding decision-making as processually immanent for the broader question of agency—and, specifically, for the normative aspects of this question, which my emphasis on the partisan and conflictual dimensions of political agency has made a necessary part of the model I have been developing. The connotations of allegiance and alignment in Ahmed's "what a body gets behind" helpfully stress this normative aspect, drawing our attention in particular to problems of motivation and responsibility. ("Why get behind *this*?" "What does it mean to have got behind *this*?") With agency distributed across the alliance, how should these dimensions be configured?

Looking back over the development of our model, we can see that my addition of an antagonistic and politically partisan quality to existing accounts of distributed agency affects this configuration of normativity in two senses. The first is retrospective: in common with these existing accounts, our model offers a descriptive picture of how agency has been distributed across an alliance and, with this, of the differential distribution among participants of capacities and local responsibility—which is where we meet the question of normativity. The difference made to this retrospective view by my focus on specifically antagonistic alliances comes, I have claimed, in the sharper delineations of responsibility this focus permits. Understanding the participants in an alliance as eminent sites of accountability offers *more and better differentiation* of their respective contributions—and, hence, of their local responsibility—by virtue of tracing these contributions in their confrontational effectivity. This, I think, is where our model addresses the second of those questions of allegiance and alignment: what does it mean to have got behind *this*? As we explored in detail in horizons 1 and 2, this means asking: for what is the alliance responsible? And what kinds and degrees of local responsibility are distributed across its participants?

The second sense in which this focus on antagonism affects the consideration of normative dimensions within the framework of distributed agency is not retrospective but prospective; and here, we come closer to the particular stakes of my attempt to understand political agency as both distributed and *decisive*. For as we have seen, the decision introduces the question of the future; and the decision as partisan engagement opens a time of responsibility understood not as retrospective attribution but as a future-oriented commitment to the promotion of *this* tendency, the prescription of *this* value. It is here, then, that our model addresses the first of those questions of allegiance and alignment: *why* get behind this? Tackling this question at the level of the alliance reminds us that the framework of distributed agency has to do without various familiar versions of purposiveness, from human-exceptionalist intentionality to the complex nonconscious relation to finality that Connolly analyzes as "teleo-dynamism" (Connolly 2013:8). Although they might be combined into an account of agency as composite, these versions of purposiveness remain a priori qualities of given types of being: they are not the form in which we encounter the committed futural orientation of the alliance. To find this form, we in fact have to combine the retrospective and the prospective and ask: what purpose will the alliance have shown, what motivation will it have expressed, what normative claim will it have made, by virtue of acting in this way? As this insistent future perfect tense indicates, the normativity of the antagonistic alliance is performative; its allegiance is given in its action. But this does not rule out consideration of motivation and projection. For in its action, the alliance crystallizes the existential stakes of its situation; its action is thickly motivated, as the conjoint expression of its participants' various concerns and interests. And in this composite action, the antagonistic alliance makes a normative claim, fighting to prescribe the value declared in this conjoint expression, composed of these various motivations and engaging the future in the mode of normative commitment. It is in this way that the action of an antagonistic alliance makes a claim on the future: as the composite enactment of a commitment to the values expressed in its decisive intervention.

In phenomenological—or better, here, quasi-phenomenological—mode, we might find ourselves wanting to look further into this question of motivation, to scrutinize in more detail the quality of experience entailed in such composite interventions. To ask, that is, some version of the quasi-phenomenological question par excellence, morphing Thomas Nagel's "What is it like to be a bat?" (Nagel 1974), through Ian Bogost's alien-phenomenological "What is it like to be a thing?" (Bogost 2012), into "What is it like to be a participant?" or even "What is it like to be an alliance?" The quasi-phenomenal character of its constitutive contributions—the mutual opacity obtaining between their locations—means that for the alliance, there is no answer to this question; there can be no synthesis of experience at the level of the alliance. Given that consciousness is a property of some participants in some alliances, however, attributed to them as a property by virtue of their differential position relative to other participants, it can be meaningful to ask what it is like to be a participant. One answer to this question follows straightforwardly from our model: conscious participants experience a share of the agency of the alliance as they emerge as its contributors, "getting behind" its interventions and motivated by their attachment to the collective existential stakes these crystallize.

If the ghost of the sovereign human individual keeps objecting to the dispersal of its sovereignty entailed in this scenario, as the commanding intentional projection of future action is replaced by the immanent inflection of processual flows, we can reply both that it was ever thus (nothing and nobody ever acts alone) and that this scenario presents less the cancellation than the displacement and relativization of that fantasy sovereignty. In some cases—such as the digital-human decision-making we considered in horizon 2—the alliance includes participants that look to the future, calculate, and contribute accordingly. And if even these local decisions must be understood here as always already processually enmeshed, as quasi-epiphenomena surfaced from ongoing flows thanks to multiple "phases of genesis" (as Bennett writes, glossing Michel Serres [2020:83]), they are not thereby exiled from any agency, as the misleading human-exceptionalist "all or nothing" would have it. Rather, they share in agency and responsibility—including in their strongly decisive, future-oriented mode—if and when they emerge *in the act*—and they do—as contributors to an antagonistic alliance, itself constituted as expressing the collective existential motivation in which they also share, whose value the alliance seeks to prescribe by its intervention.

If the emphasis of our model on partisan commitment compels it to consider these normative dimensions, does the model itself have normative force? In arguing for this way of understanding political agency, am I myself seeking to prescribe a particular value or set of values? Philosophical respectability would require me to answer quickly in the negative: it is, of course, generally accepted as axiomatic that prescriptions, or values, cannot be derived from descriptions, or facts. And I have indeed sought throughout this study to formulate my analyses in such a way that they might equally well describe the operation of alliances with drastically different orientations. As we saw in horizon 1, the same beings can emerge as participants in opposing alliances; and the model developed here functions, I believe, to account for the actions of an antagonistic agential alliance regardless of its particular commitment. Moreover, our model will account for the actions of an alliance whether or not that alliance itself accepts the ontological premises of the model. And this is where that respectably philosophical axiomatic separation of ontological description from ethical or political prescription starts to break down. For it is no less axiomatic that ontological descriptions encode normative commitments. To refuse the ontological premises of our model is to align oneself with a position that itself makes a normative claim: namely, that we *should* understand the world through a substance-metaphysical hierarchy of capacities. And, of course, to start, as I have done, from a commitment to ontological egalitarianism is both to oppose that other position and to affirm the value of understanding the world through this framework.

In this sense, my descriptions do, indeed, seek to be normatively prescriptive. In modeling action in this way, I want to persuade you that it is not only plausible but desirable to adopt this approach. This is the interpretation I am putting forth and to which this book will have expressed my commitment: that in the current planetary predicament, it is necessary both to reject human exceptionalism and to retain trenchant political effectivity; that as a consequence, we need a way of understanding political agency as both distributed and decisive; and that the model developed here might contribute to such an understanding. To recall Latour's terminology, if I certainly believe that this model accounts equally well for the partisan political interventions of Humans and Earthlings, only an Earthling would propose such a model of political agency in the first place.

If I am proposing this model, then, it is in the belief and hope that it might be helpful now, that its reconciliation of distributed agency and thickly motivated, decisive partisan intervention might shift the philosophical-theoretical stand-off between human exceptionalists and ontological egalitarians and, in so doing, enrich the efforts of Earthling alliances by integrating into their commitment to plurality something of the sharp effectivity the exceptionalists claim as exclusively theirs. For in our understanding of political action, the stakes are indeed those of this Human-Earthling conflict played out through its various composite alliances.

Granted, the model proposed here might not look much like politics as we have known it, or as our human dreams of exclusive agency let us imagine it. Its active principle is neither the autonomous, rational liberal subject nor the solely human collective, both of which act through decisions that might be more or less contextually determined but that remain for their authors transparently the products of their will. This much is already given by our exit from the theater of intentionality, by virtue of which we are obliged to reformulate political agency in terms of the composition and emergence that define agency freed from the transcendent projections we have always tied to self-representation. And in terms of our model of politics, it is not just self-representation but representation in general that must be drastically reconfigured. With politics no longer the activity of humans alone, we are no longer confined to discursive mediation, to the resonance of the *logos* in the agora; nor are we concerned only with the self-identification and dialectical recognition of human collectives bound together in their struggle to affirm their respective vital interests. In the expanded field we here encounter, these interests are enmeshed with those of beings of all kinds, and if the affirmation of these interests in the action of the alliance constitutes, for this very reason, a decisive intervention into its situation, this no longer redistributes a solely human sensible realm. The claim made by the composite alliance is not a Rancierian claim to recognition; its participants are not insisting that they, too, be counted at last as bearers of the qualifying linguistic-cognitive capacity—that what had been abusively heard only as their *noise* be now admitted as *speech* (see Rancière 2004; 2009). The claim is indeed that these interests should be taken into account—but this claim is not represented, it is performed, affirmed as the alliance's motivating value in and by its combative action.

As we have seen, various elements of human-exceptionalist politics as representation do persist in our model: there are fora, there is language, and, indeed, there is representation. But these are now relativized as local contributions in a wider struggle: namely—as we saw in the introduction—the conflict over "the modalities of life in common" (Marchetti 2013:14) or the Latourian composition of the common world (see, for example, Latour 2014:15–16), in a drastically expanded sense: with many more participants and, in consequence, greatly intensified stakes. And, of course, the key element of human-exceptionalist models of politics that we have retained here is the possibility of decisive, committed intervention. That agency emerges only as the property of a composite alliance means that this intervention must now be understood as processually immanent; but thanks to Stiegler and Malabou, we have been able to understand such immanent strong discontinuity as inflection and interruption together—a watershed, or a fold with the force of a cut. An "agential cut," in that key image from Barad (1997:140), in which the elements of a situation emerge, in the act, in the radically new configuration that is the alliance, and the situation is divided, traversed by the agential front along which the alliance pursues its conflict.

In this respect, the immanent strong discontinuity of agential emergence bears comparison with the Event as conceptualized by Badiou, that irruption into a situation of the element that could not figure within it, after which the situation is irrevocably changed. What is more, in this context, the retroactive temporality of Stiegler's Deleuzian quasi-cause and Malabou's plastic reading, which have defined for us the futural partisan commitment of the emergent agential alliance, calls to mind the emergence with the Event of Badiou's Subject, defined by its commitment (in Badiou's terms, its fidelity) to the value (for Badiou: the Truth) of this interruption (see especially Badiou 2005a); and the normative commitment performed in this emergence likewise calls to mind Badiou's insistence on the irreducible prescriptive dimension of politics, as seen in the introduction (see esp. Badiou 2005b).

As I mentioned in the introduction, I am well aware that Badiou's assertive human exceptionalism makes this *rapprochement* problematic, and I would certainly not want to push this comparison too far. Not only this human exceptionalism but the related elevation of the Subject and the more or less messianic, transcendent status of the Event would plainly make this a very short venture, of limited interest at best. But if I have invoked Badiou here, at the end of a book whose commitments are in important ways plainly not his, it is to signal by his very incompatibility with my motivating ontological egalitarianism a key part of what I hope the model developed here might contribute as an understanding of politics. Although it would be excessive to define political agency, let alone agency as a whole, solely or primarily in terms of partisan conflict, my contention here has been that such conflict must feature in any overall account of politics, and I have argued that it is indeed compatible with an account of political agency as distributed agency. The combative, prescriptive dimension that is politics for Badiou has doubtless here played a less definitional role, but it is no less irreducible.

I suggested in the introduction that we might think of the accident as a less grandiose version of Badiou's Event. That suggestion should by now carry more substantive force than it did all those pages ago. With Stiegler and Malabou, we have come to understand the accident as the locus of agential emergence, naming in its wildness the void that any consistent conception of emergence must needs substitute for the determining origin of substance metaphysics, and the quasi-causal transformation of this void into the site of the decisive commitment in which the agent emerges. And there is no agent here other than plural, composite: it is the alliance that acts, accidentally, transformatively, decisively emergent all along its composite agential front—what Emily Apter calls "a skein of 'eventalness' " (2019). The agential alliance comes together *in the act*, the conjunction of its participants' contributions; and as the emergence of a new formation, this conjunction both reconfigures its situation and is determined by nothing in advance. As a decisive, committed, prescriptive intervention, it is also, in this sense, an accident. *Accidental* political agency, then. Yes, this is a kind of overstatement: the element of futural commitment essential to our model distances it from the hapless stumbling we are compelled to see in this figure by our persisting addiction to fantasies of sovereign control. Only from the shores of these fantasies does the accident look so derelict, however. And nothing obliges us to adopt this point of view, especially if, as I have argued, accidental agential emergence is indeed compatible with effective, *decisive* action. What this gives us, then, is distributed political agency as contingent, composite frontal engagement: emergent, accidental, immune to the capital Event but effecting decisive, partisan intervention, combatively promoting *this* tendency, prescribing *this* value.

As the demonstrators' chant has it, "Tell me what democracy looks like—*this* is what democracy looks like!" Or, rather, if we are committed to ontological egalitarianism and want a fully convincing account of political effectivity, *this* is what *politics* looks like. All around us, meeting in such frontal engagement, are composite alliances defined by their respective commitments to fundamentally incompossible futures. On one side, the exceptionalist delusions and brutal politics of class, race, and species sacrifice underpinning the diminishing returns of extractivism and the interstellar fantasies of the Human oligarchs, determined to mistake their allies and conscripts for mere instruments. On the other, Earthling alliances mobilizing through and in the name of their own variegated composition. A confrontation concerning the nature and quality of life in common: a political conflict. And in this conflict, the political agency we have traced through this study, distributed and decisive, composite and confrontational, is already at work.

WORKS CITED

Ahmed, Sara. 2014. *Willful Subjects*. Durham, NC: Duke University Press.

Alston, Philip. 2019. "World Stumbling Zombie-Like Into a Digital Welfare Dystopia, Warns UN Human Rights Expert." Report of the special rapporteur on extreme poverty and human rights, presented to the United Nations General Assembly, October 18, 2019. https://www.ohchr.org/EN/NewsEvents/Pages/DisplayNews.aspx?NewsID=25156.

Amoore, Louise. 2009. "Algorithmic War: Everyday Geographies of the War on Terror." *Antipode* 41, no. 1: 49–69.

——. 2020. *Cloud Ethics: Algorithms and the Attributes of Ourselves and Others*. Durham, NC: Duke University Press.

Amoore, Louise, and Rita Raley. 2017. "Securing with Algorithms: Knowledge, Decision, Security." *Security Dialogue* 48, no. 1: 3–10.

Appadurai, Arjun. 2015. "Mediants, Materiality, Normativity." *Public Culture* 27, no. 2: 221–37.

Apter, Emily. 2018. *Unexceptional Politics: On Obstruction, Impasse, and the Impolitic*. London: Verso.

——. 2019. "Blurring the Event: Micropolitics and Ecosophy, May '68 to the ZAD." Talk given at the Institute for Cultural Inquiry, Berlin, March 11, 2019. https://www.ici-berlin.org/events/emily-apter/.

Auletta, Gennaro. 2011. *Cognitive Biology: Dealing with Information from Bacteria to Minds*. Oxford: Oxford University Press.

Bacon, Francis. 1902. *Novum Organum, or True Suggestions for the Interpretation of Nature*, ed. Joseph Devey. New York: P. F. Collier and Son. http://www.gutenberg.org/files/45988/45988-h/45988-h.htm. Accessed July 8, 2021.

Badiou, Alain. 2005a. *Being and Event*, trans. Oliver Feltham. London: Continuum.

——. 2005b. *Metapolitics*, trans. Jason Barker. London: Verso.

——. 2009. *Logics of Worlds: Being and Event, 2*, trans. Alberto Toscano. London: Continuum.

Balibar, Étienne, and Sandra Laugier. 2014. "Agency." In *Dictionary of Untranslatables: A Philosophical Lexicon*, ed. Barbara Cassin, trans. Steven Rendall et al. Princeton: Princeton University Press.

Barad, Karen. 1997. *Meeting the Universe Halfway: Quantum Physics and the Entanglement of Matter and Meaning*. Durham, NC: Duke University Press.

Barker, Stephen. 2013. "Techno-Pharmaco-Genealogy." In *Stiegler and Technics*, ed. Christina Howells and Gerald Moore, 259–75. Edinburgh: Edinburgh University Press.

Barto, E. Kathryn, Jeffrey D. Weidenhamer, Don Cipollini, and Matthias C. Rillig. 2012. "Fungal Superhighways: Do Common Micorrhizal Networks Enhane Below Ground Communication?" *Trends in Plant Science* 17, no. 11: 633–37.

Bateson, Gregory. 1972. *Steps to an Ecology of Mind*. London: Intertext.

Beardsworth, Richard. 1998. "Thinking Technicity." *Cultural Values* 2, no. 1: 70–86.

Bennett, Jane. 2010. *Vibrant Matter: A Political Ecology of Things*. Durham, NC: Duke University Press.

——. 2020. *Influx and Efflux: Writing Up with Walt Whitman*. Durham, NC: Duke University Press.

Bennett, W. Lance, and Alexandra Segerberg. 2015. "The Logic of Connective Action: Digital Media and the Personalization of Contentious Politics." In *The Handbook of Digital Politics*, ed. Stephen Coleman and Deen Freelon, 169–98. Cheltenham, UK: Edward Elgar.

Bennington, Geoffrey. 1996. "Emergencies." *Oxford Literary Review* 18, no. 1-2: 175–216.

Berardi, Franco "Bifo." 2019. "Prophecy and Computation." Verso (blog), March 26, 2019. https://www.versobooks.com/blogs/4274-prophecy-and-computation?utm_source=Master+List&utm_campaign=09cc2346dd-US+Direct%3A+Jacobin+bookshelf_COPY_01&utm_medium=email&utm_term=0_1f96ba5fab-09cc2346dd-410285045&mc_cid=09cc2346dd&mc_eid=3e07383d41.

Bergthaller, Hannes. 2014. "Limits of Agency: Notes on the Material Turn from a Systems-Theoretical Perspective." In *Material Ecocriticism*, ed. Serenella Iovino and Serpil Opperman. Bloomington: Indiana University Press.

Bineau, Manon, and Antoine Chopot. 2017. "*Nous sommes la nature qui se défend contre l'économie.*" ("We are nature defending itself against the economy.") *ÉcoRev': Revue critique d'écologie politique* 44: 43–51.

Blok, Anders, and Torben Elgaard Jensen. 2011. *Bruno Latour: Hybrid Thoughts in a Hybrid World*. London: Routledge.

Bogost, Ian. 2012. *Alien Phenomenology: What It's Like to Be a Thing*. Minneapolis: University of Minnesota Press.

Bolin, Göran, and Jonas Andersson Schwarz. 2015. "Heuristics of the Algorithm: Big Data, User Interpretation and Institutional Translation." *Big Data & Society* July–December: 1–12.

Bostrom, Nick. 2014. *Superintelligence: Paths, Dangers, Strategies*. Oxford: Oxford University Press.

Bratton, Benjamin H. 2010. "Surviving the Interface: The Envelopes, Membranes and Borders of Deep Cosmopolitics" (video). October 2010. https://web.archive.org/web/20131221051525/http://www.bratton.info/projects/talks/surviving-the-interface/. Accessed July 8, 2021.

——. 2015. *The Stack*. Cambridge, MA: MIT Press.

——. 2019. *The Terraforming*. Moscow: Strelka.

Bridle, James. 2018. *New Dark Age: Technology and the End of the Future*. London: Verso.

Brown, Wendy, Peter E. Gordon, and Max Pensky. 2018. *Authoritarianism: Three Inquiries in Critical Theory*. Chicago: Chicago University Press.

Buchanan, Ian. 2015. "Assemblage Theory and Its Discontents." *Deleuze Studies* 9, no. 3: 382–92.

Burns, Michael O'Neill. 2016. "Vitality or Weakness?" *Angelaki* 21, no. 4: 11–22.

Callon, Michel, and John Law. 1995. "Agency and the Hybrid *Collectif*." *South Atlantic Quarterly* 94: 481–507.

Callon, Michel, Pierre Lascoumes, and Yannick Barthe. 2009. *Acting in an Uncertain World: An Essay on Technical Democracy*, trans. Graham Burchell. Cambridge, MA: MIT Press.

Caygill, Howard. 2013. *On Resistance: A Philosophy of Defiance*. London: Bloomsbury Academic.

Chakrabarty, Dipesh. 2009. "The Climate of History: Four Theses." *Critical Inquiry* 35: 197–222.

Chalmers, David J. 2008. "Strong and Weak Emergence." In *The Re-Emergence of Emergence*, ed. Phillip Clayton and Paul Davies, 244–54. Oxford: Oxford University Press.

Citi GPS. 2016. "Technology at Work v 2.0: The Future Is Not What It Used to Be." https://www.oxfordmartin.ox.ac.uk/publications/technology-at-work-v2-0-the-future-is-not-what-it-used-to-be/. January 27, 2016.

Clark, Andy, and David J. Chalmers. 1998. "The Extended Mind." *Analysis* 58: 7–19.

Clark, Nigel, and Kathryn Yusoff. 2017. "Geosocial Formations and the Anthropocene." *Theory, Culture & Society* 34, nos. 2–3: 3–23.

Clayton, Phillip. 2008. "Conceptual Foundations of Emergence Theory." In *The Re-Emergence of Emergence*, ed. Phillip Clayton and Paul Davies, 1–32. Oxford: Oxford University Press.

Clough, Patricia Ticineto. 2018. *The User Unconscious: On Affect, Media, and Measure*. Minneapolis: University of Minnesota Press.

Coeckelbergh, Mark. 2015. *Money Machines: Electronic Financial Technologies, Distancing, and Responsibility in Global Finance*. Farnham, UK: Ashgate.

Cole, David. 2019. "The Path of Greatest Resistance." *New York Review of Books*, 21–22. February 7, 2019.

Colebrook, Claire. 2014. *Death of the PostHuman: Essays on Extinction*. Ann Arbor: Open Humanities Press. http://www.openhumanitiespress.org/books/titles/death-of-the-posthuman/.

——. 2017. "Impossible, Unprincipled, and Contingent: Bernard Stiegler's Project of Revolution and Redemption." *Boundary 2* 44, no. 1: 213–37.

Colony, Tracy. 2011. "Epimetheus Bound: Stiegler on Derrida, Life, and the Technological Condition." *Research in Phenomenology* 41: 72–89.

Connolly, William E. 2010. *A World of Becoming*. Durham, NC: Duke University Press.

——. 2013. *The Fragility of Things: Self-Organizing Processes, Neoliberal Fantasies, and Democratic Activism*. Durham, NC: Duke University Press.

——. 2017. *Facing the Planetary: Entangled Humanism and the Politics of Swarming*. Durham, NC: Duke University Press.

Coole, Diana. 2013. "Agentic Capacities and Capacious Historical Materialism: Thinking with New Materialisms in the Political Sciences." *Millennium: Journal of International Studies* 41, no. 3: 451–69.

Crowley, Martin. 2019. "The Many Worlds of Jean-Luc Nancy." *Paragraph* 42, no. 1: 22–36.

Danaher, John. 2016. "The Threat of Algocracy: Reality, Resistance and Accommodation." *Philosophy and Technology* 29, no. 3: 245–68.

Danowski, Déborah, and Eduardo Viveiros de Castro. 2017. *The Ends of the World*, trans. Rodrigo Nunes. Cambridge: Polity.

Debaise, Didier. 2017. *Nature as Event: The Lure of the Possible*, trans. Michael Halewood. Durham, NC: Duke University Press.

Debaise, Didier, and Isabelle Stengers. 2017. "The Insistence of Possibles: Towards a Speculative Pragmatism" (trans. Angela Brewer). *Parse* 7: 12–19.

De Beistegui, Miguel. 2013. "The New Critique of Political Economy." In *Stiegler and Technics*, ed. Christina Howells and Gerald Moore, 181–91. Edinburgh: Edinburgh University Press.

DeLanda, Manuel. 2006. *A New Philosophy of Society: Assemblage Theory and Social Complexity*. London: Continuum.

Deleuze, Gilles. 1989. *Cinema 2: The Time-Image*, trans. Hugh Tomlinson and Robert Galeta. Minneapolis: University of Minnesota Press.

——. 1990. *The Logic of Sense*, trans. Mark Lester with Charles Stivale. London: Athlone.

——. 1992. "Postscript on the Societies of Control." *October* 59: 3–7.

Deleuze, Gilles, and Félix Guattari. 2014. *A Thousand Plateaus: Capitalism and Schizophrenia*, trans. Brian Massumi. London: Bloomsbury Academic.

Derrida, Jacques. 1976. *Of Grammatology*, trans. Gayatri Chakravorty Spivak. Baltimore: Johns Hopkins University Press.

——. 1981. *Dissemination*, trans. Barbara Johnson. London: Athlone.

——. 1990. "Force of Law: The 'Mystical Foundation of Authority,' " trans. Mary Quaintance. *Cardozo Law Review* 11: 919–1045.

——. 2006. *Specters of Marx: The State of the Debt, the Work of Mourning and the New International*, trans. Peggy Kamuf. London: Routledge.

Derrida, Jacques, and Bernard Stiegler. 2002. *Echographies of Television: Filmed Interviews*, trans. Jennifer Bajorek. Cambridge: Polity.

Descartes, René. 1649. *A Discourse of a Method for the Well Guiding of Reason, and the Discovery of Truth in the Sciences*. London: Thomas Newcombe. https://www.gutenberg.org/files/25830/25830-h/25830-h.htm. Accessed August 13, 2021.

De Vries, Gerard. 2016. *Bruno Latour*. Cambridge: Polity.

Diderot, Denis. 1984a. "*Droit naturel*." ("Natural Law.") In *L'Encyclopédie, ou Dictionnaire raisonné des sciences, des arts et des métiers: Textes choisis*, ed. Philippe Goujard, 141–47. Paris: Messidor/Éditions Sociales.

——. 1984b. "*Homme*." ("Man.") In *L'Encyclopédie, ou Dictionnaire raisonné des sciences, des arts et des métiers: Textes choisis*, ed. Philippe Goujard, 225–30. Paris: Messidor/Éditions Sociales.

Dieter, Michael, and David Gauthier. 2019. "On the Politics of Chrono-Design: Capture, Time and the Interface." *Theory, Culture & Society* 36, no. 2: 61–87.

Dixon-Román, Ezekiel. 2016. "Algo-Ritmo: More-Than-Human Performative Acts and the Racializing Assemblages of Algorithmic Architectures." *Cultural Studies ↔ Critical Methodologies* 16, no. 5: 482–90.

Dresp-Langley, Birgitta. 2012. "Why the Brain Knows More Than We Do: Non-Conscious Representations and Their Role in the Construction of Conscious Experience." *Brain Science* 2, no. 1: 1–12.

Ekman, Ulrik. 2007. "Of Transductive Speed—Stiegler." *Parallax* 13, no. 4: 46–63.

Elam, Mark. 1999. "Living Dangerously with Bruno Latour in a Hybrid World." *Theory, Culture & Society* 16, no. 4: 1–24.

Elder-Vass, Dave. 2010. *The Causal Power of Social Structures: Emergence, Structure and Agency*. Cambridge: Cambridge University Press.

Eubanks, Virginia. 2018. *Automating Inequality: How High-Tech Tools Profile, Police, and Punish the Poor*. New York: St Martin's.

Floridi, Luciano, and Sanders, J. W. 2004. "On the Morality of Artificial Agents." *Minds and Machines* 14, no. 3: 349–79.

Foucault, Michel. 1978. *The History of Sexuality, 1: An Introduction*, trans. Robert Hurley. New York: Pantheon.

——. 2003. *Society Must Be Defended: Lectures at the Collège de France 1975–1976*, trans. David Macey. New York: Picador.

——. 2007. *Security, Territory Population. Lectures at the Collège de France 1977–1978*, trans. Graham Burchell. Basingstoke, UK: Macmillan.

Franke, Anselm. 2014. "The Forensic Scenography." In *Forensis: The Architecture of Public Truth*, ed. Forensic Architecture, 483–94. Berlin: Sternberg Press & Forensic Architecture.

Frey, Carl Benedikt, and Michael A. Osborne. 2013. "The Future of Employment: How Susceptible are Jobs to Computerisation?" September 17, 2013. http://www.oxfordmartin.ox.ac.uk/downloads/academic/The_Future_of_Employment.pdf.

Fuchs, Christian. 2016. *Reading Marx in the Information Age: A Media and Information Studies Perspective on "Capital Volume 1."* London: Routledge.

Galloway, Alexander R. 2020. "Mathification." In "Economies of Existence," ed. Emily Apter and Martin Crowley (special issue). *Diacritics* 47, no. 1): 95–115.

Galloway, Alexander R., and Eugene Thacker. 2007. *The Exploit: A Theory of Networks*. Minneapolis: University of Minnesota Press.

Gerbaudo, Paolo. 2019. *The Digital Party: Political Organisation and Online Democracy*. London: Pluto.

Gibson, Andrew. 2012. *Intermittency: The Concept of Historical Reason in Recent French Philosophy*. Edinburgh: Edinburgh University Press.

Giovannetti, Manuela, Luciano Avio, Paola Fortuna, Elisa Pellegrino, Cristiana Sbrana, and Patrizia Strani. 2006. "At the Root of the Wood Wide Web." *Plant Signaling & Behavior* 1, no. 1: 1–5.

Gratton, Peter. 2014. *Speculative Realism: Problems and Prospects*. London: Bloomsbury. Greenfield, Adam. 2018. *Radical Technologies: The Design of Everyday Life*. London: Verso.

Gregory, Karen. n.d. "Weird Solidarities." *DIS Magazine: The Data Issue*. http://dismagazine.com/discussion/72958/karen-gregory-weird-solidarities/. Accessed July 8, 2021.

Grove, Jairus. 2015. "Something Darkly This Way Comes: The Horror of Plasticity in an Age of Control." In *Plastic Materialities: Politics, Legality, and Metamorphosis in the Work of Catherine Malabou*, ed. Brenna Bhandar and Jonathan Goldberg-Hiller, 233–63. Durham, NC: Duke University Press.

Hallward, Peter. 2006. *Out of This World: Deleuze and the Philosophy of Creation*. London: Verso.

Hansen, Mark B.N. 2015. *Feed-Forward: On the Future of Twenty-First-Century Media*. Chicago: University of Chicago Press.

——. 2016. "Topology of Sensibility." In *Ubiquitous Computing, Complexity, and Culture*, ed. Ulrik Ekman et al., 80–106. London: Routledge. Online edition.

Haraway, Donna J. 1992. "The Promises of Monsters: A Regenerated Politics for Inappropriate/d Others." In *Cultural Studies*, ed. Lawrence Grossberg, Cary Nelson, and Paula A. Treichler, 295–337. New York; Routledge.

——. 1997. *Modest_Witness@Second_Millenium.FemaleMan[C]_Meets_Onco -Mou seTM: Feminism and Technoscience*. New York: Routledge.

——. 2006. "When We Have Never Been Human, What Is to Be Done?" (interview by Nicholas Gane.) *Theory, Culture & Society* 23, no. 7-8: 135–58.

——. 2016. *Staying with the Trouble: Making Kin in the Chthulucene*. Durham, NC: Duke University Press.

Harcourt, Bernard E. 2015. *Exposed: Desire and Disobedience in the Digital Age*. Cambridge, MA: MIT Press.

Harding, Sandra. 2008. *Sciences from Below: Feminisms, Postcolonialities, and Modernities*. Durham, NC: Duke University Press.

Harman, Graham. 2009. *Prince of Networks: Bruno Latour and Metaphysics*. Melbourne: Re.Press.

——. 2014. *Bruno Latour: Reassembling the Political*. London: Pluto.

Haworth, Michael. 2016. "Bernard Stiegler on Transgenerational Memory and the Dual Origin of the Human." *Theory, Culture & Society* 33, no. 3: 151–73.

Hayles, N. Katherine. 2012. *How We Think: Digital Media and Contemporary Technogenesis*. Chicago: Chicago University Press.

——. 2017. *Unthought: The Power of the Cognitive Nonconscious*. Chicago: University of Chicago Press.

Heidegger, Martin. 1993. *Basic Writings*, ed. David Farrell Krell. London: Routledge.

——. 1995. *The Fundamental Concepts of Metaphysics: World, Finitude, Solitude*, trans. William MacNeill and Nicholas Walker. Bloomington: Indiana University Press.

Heller Charles, and Lorenzo Pezzani. 2014. "Liquid Traces: Investigating the Deaths of Migrants at the EU's Maritime Frontier." In *Forensis: The Architecture of Public Truth*, ed. Forensic Architecture, 657–84. Berlin: Sternberg Press & Forensic Architecture.

Hope, Alexander. 2014. "The Future Is Plastic: Refiguring Malabou's Plasticity." *Journal for Cultural Research* 18, no. 4: 329–49.

Hornborg, Alf. 2017. "Artifacts Have Consequences, Not Agency: Toward a Critical Theory of Global Environmental History." *European Journal of Social Theory* 20, no. 1: 95–110.

Ingold, Tim. 2011. *Being Alive: Essays on Movement, Knowledge, Description*. Abingdon, UK: Routledge.

——. 2016. "On Human Correspondence." *Journal of the Royal Anthropological Institute* 23: 9–27.

Invisible Committee. 2015. *To Our Friends*, trans. Robert Hurley. South Pasadena, CA: Semiotext(e).

James, Ian. 2012. *The New French Philosophy*. Cambridge: Polity.

——. 2019. *The Technique of Thought: Nancy, Laruelle, Malabou and Stiegler After Naturalism*. Minneapolis: University of Minnesota Press.

Johnson, Christopher. 2013. "The Prehistory of Technology: On the Contribution of Leroi-Gourhan." In *Stiegler and Technics*, ed. Christina Howells and Gerald Moore, 34–52. Edinburgh: Edinburgh University Press.

Johnston, Adrian. 2014. *Adventures in Transcendental Materialism: Dialogues with Contemporary Thinkers*. Edinburgh: Edinburgh University Press.

Kant, Immanuel. 1917. *Perpetual Peace: A Philosophical Essay*, trans. M. Campbell Smith. London: George Allen & Unwin. Project Gutenberg: http://www.gutenberg.org/ebooks/50922. Accessed July 8, 2021.

——. 1998. *Critique of Pure Reason*, trans. and ed. Paul Guyer and Allen W. Wood. Cambridge: Cambridge University Press.

——. 2012. *Anthropology from a Pragmatic Point of View*, trans. and ed. Robert B. Louden. Cambridge: Cambridge University Press.

Karpf, David. 2017. *Analytic Activism: Digital Listening and the New Political Strategy*. Oxford: Oxford University Press.

Kavelakis, Stathis. 2018. "Borderland: Greece and the EU's Southern Question." *New Left Review* 110: 5–33.

Keucheyan, Razmig. 2016. *Nature Is a Battlefield*, trans. David Broder. Cambridge: Polity.

Kirby, Vicki. 2016. "Grammatology: A Vital Science." *Derrida Today* 9, no. 1: 47–67.

Klein, Naomi. 2007. *The Shock Doctrine: The Rise of Disaster Capitalism*. London: Allen Lane.

——. 2014. *This Changes Everything: Capitalism vs. the Climate*. London: Allen Lane.

Kohn, Eduardo. 2013. *How Forests Think: Toward an Anthropology Beyond the Human*. Berkeley: University of California Press.

Koopman, Colin. 2019. *How We Became Our Data: A Genealogy of the Informational Person*. Chicago: University of Chicago Press.

Kuitenbrouwer, Klass. 2015. "The Stack and the Post-Human User: An Interview with Benjamin Bratton." https://tuinvanmachines.hetnieuweinstituut.nl/en/stack-and-posthuman-user-interview-benjamin-bratton. Accessed July 8, 2021.

Lahoud, Adrian. 2014. "Floating Bodies." In *Forensis: The Architecture of Public Truth*, ed. Forensic Architecture, 495–518 Berlin: Sternberg Press & Forensic Architecture.

Laruelle, François. 2013. *Dictionary of Non-Philosophy*, trans. Taylor Adkins. Minneapolis: Univocal.

Latour, Bruno. 1988. *The Pasteurization of France*, trans. Alan Sheridan and John Law. Cambridge, MA: Harvard University Press.

——. 1993. *We Have Never Been Modern*, trans. Catherine Porter. Cambridge, MA: Harvard University Press.

——. 1998. "*Ein Ding ist Ein Thing*: A (Philosophical) Programme for a Left (European) Party." *Concepts and Transformation* 3, no. 1-2: 97–112.

——. 1999. *Pandora's Hope: Essays on the Reality of Science Studies*. Cambridge, MA: Harvard University Press.

——. 2002. "Si l'on parlait un peu politique?" ("Shall we talk a little politics?") *Politix* 15, no. 58: 143–65.

——. 2004. *Politics of Nature: How to Bring the Sciences Into Democracy*, trans. Catherine Porter. Cambridge, MA: Harvard University Press.

——. 2005. *Reassembling the Social: An Introduction to Actor-Network-Theory*. Oxford: Oxford University Press.

——. 2007. "A Plea for the Earthly Sciences." In *New Social Connections: Sociology's Subjects and Objects*, ed. Judith Burnett, Syd Jeffers, and Graham Thomas, 72–84. London: Palgrave Macmillan.

——. 2012a. *An Inquiry Into Modes of Existence: An Anthropology of the Moderns*. Modes of Existence. Augmented online version. http://www.modesofexistence.org. Accessed July 8, 2021.

——. 2012b. "*L'Oeuvre de Bruno Latour: Une pensée politique exégétique.*" ("The work of Bruno Latour: An exegetical form of political thought.") (interview by Laurent Godmer and David Smadja.) *Raisons politiques* 47: 115–48.

——. 2013. *An Inquiry Into Modes of Existence: An Anthropology of the Moderns*, trans. Catherine Porter. Cambridge, MA: Harvard University Press.

——. 2014. "Agency at the Time of the Anthropocene." *New Literary History* 45: 1–18.

——. 2015. "Telling Friends from Foes in the Time of the Anthropocene." In *The Anthropocene and the Global Environment Crisis: Rethinking Modernity in a New Epoch*, ed. Clive Hamilton, Christophe Bonneuil, and François Gemenne. London: Routledge. Online edition.

——. 2017. *Facing Gaia: Eight Lectures on the New Climatic Regime*, trans. Catherine Porter. Cambridge: Polity.

——. 2018. *Down to Earth: Politics in the New Climatic Regime*, trans. Catherine Porter. Cambridge: Polity.

——. 2019. "*Du bon usage de la consultation nationale.*" ("On the proper use of national consultation.") AOC, January 14, 2019. https://aoc.media/analyse/2019/01/14/usage-de-consultation-nationale.

LaVaque-Manty, Mika. 2002. *Arguments and Fists: Political Agency and Justification in Liberal Theory*. London: Routledge.

Lemke, Thomas. 2018. "An Alternative Model of Politics? Prospects and Problems of Jane Bennett's *Vital Materialism*." *Theory, Culture & Society* 35, no. 6: 31–54.

Lewis, Michael. 2013. "Of a Mythical Philosophical Anthropology: The Transcendental and the Empirical in *Technics and Time*." In *Stiegler and Technics*, ed. Christina Howells and Gerald Moore, 53–68. Edinburgh: Edinburgh University Press.

Loewenstein, Antony. 2015. *Disaster Capitalism: Making a Killing out of Catastrophe*. London: Verso.

Luckhurst, Roger. 2006. "Bruno Latour's Scientifiction: Networks, Assemblages, and Tangled Objects." *Science Fiction Studies* 33, no. 1: 4–17.

Macfarlane, Robert. 2016. "The Secrets of the Wood Wide Web." *New Yorker*. August 7, 2016. https://www.newyorker.com/tech/annals-of-technology/the-secrets-of-the-wood-wide-web.

MacKenzie, Donald. 2019. "Just How Fast?" *London Review of Books* 41, no. 5 (March 7). https://www.lrb.co.uk/v41/n05/donald-mackenzie/just-how-fast.

MacLeod, Norman. 2016. "Response to Catherine Malabou, 'One Life Only: Biological Resistance, Political Resistance.' " *Critical Inquiry* 43, no. 1: 191–99.

Malabou, Catherine. 2005. *The Future of Hegel*, trans. Lisabeth During. Abingdon, UK: Routledge.

——. 2008a. "A Conversation with Catherine Malabou" (interview with Noelle Vahanian). *Journal for Cultural and Religious Theory* 9, no. 1: 1–13.

——. 2008b. *What Should We Do With Our Brain?*, trans. Sebastian Rand. New York: Fordham University Press.

——. 2009. *La Chambre du milieu: De Hegel aux neurosciences*. ("The middle chamber: From Hegel to the neurosciences.") Paris: Hermann.

——. 2010. *Plasticity at the Dusk of Writing*, trans. Carolyn Shread. New York: Columbia University Press.

——. 2011a. *The Heidegger Change: On the Fantastic in Philosophy*, trans. Peter Skafish. Albany, NY: State University of New York Press.

——. 2011b. *Que faire de notre cerveau?* 2nd ed. Paris: Bayard.

——. 2012a. *L'Avenir de Hegel: Plasticité, temporalité, dialectique*. 2nd ed. Paris: Vrin.

——. 2012b. *The New Wounded: From Neurosis to Brain Damage*, trans. Steven Miller. New York: Fordham University Press.

——. 2013. "Go Wonder: Subjectivity and Affects in Neurobiological Times." In *Self and Emotional Life: Philosophy, Psychoanalysis, and Neuroscience*, ed. Adrian Johnston and Catherine Malabou, 3–72. New York: Columbia University Press.

——. 2015a. "Anthropocene, a New History?" Lecture to the European Graduate School, September 20, 2015. https://www.youtube.com/watch?v=eDdTqr-5APg.

——. 2015b. *Ontology of the Accident: An Essay on Destructive Plasticity*, trans. Carolyn Shread. Cambridge: Polity.

——. 2015c. "Whither Materialism? Althusser/Darwin." In *Plastic Materialities: Politics, Legality, and Metamorphosis in the Work of Catherine Malabou*, ed. Brenna Bhandar and Jonathan Goldberg-Hiller, 48–60. Durham, NC: Duke University Press.

——. 2015d. "Will Sovereignty Ever Be Deconstructed?" In *Plastic Materialities: Politics, Legality, and Metamorphosis in the Work of Catherine Malabou*, ed. Brenna Bhandar and Jonathan Goldberg-Hiller, 35–46. Durham, NC: Duke University Press.

——. 2016a. "Before and Above: Spinoza and Symbolic Necessity." *Critical Inquiry* 43, no. 1: 84–109.

——. 2016b. *Before Tomorrow: Epigenesis and Rationality*, trans. Carolyn Shread. Cambridge: Polity.

——. 2016c. "One Life Only: Biological Resistance, Political Resistance" (trans. Carolyn Shread). *Critical Inquiry* 42, no. 3: 429–38.

——. 2018. "Is Science the Subject of Philosophy? Miller, Badiou, and Derrida Respond" (trans. William Samson). In *Thinking Catherine Malabou: Passionate Detachments*, ed. Thomas Wormald and Isabell Dahms. London: Rowman & Littlefield.

——. 2019a. *Morphing Intelligence: From IQ Measurement to Artificial Brains*, trans. Carolyn Shread. New York: Columbia University Press.

——. 2019b. "What Should We Do with Plasticity? An Interview with Catherine Malabou" (interview by Benjamin Dalton). *Paragraph* 42, no. 2: 238–54.

Malm, Andreas. 2015. "The Anthropocene Myth." Jacobin, March 30, 2015. https://www.jacobinmag.com/2015/03/anthropocene-capitalism-climate-change/.

——. 2016. *Fossil Capital: The Rise of Steam Power and the Roots of Global Warming*. London: Verso.

——. 2018. *The Progress of This Storm: Nature and Society in a Warming World*. London: Verso.

Marchart, Oliver. 2018. *Thinking Antagonism: Political Ontology After Laclau*. Edinburgh: Edinburgh University Press.

Marchetti, Raffaele. 2013. "Introduction: Political Agency in the Age of Globalization." In *Contemporary Political Agency: Theory and Practice*, ed. Brice Maiguashca and Raffaele Marchetti, 12–29. London: Routledge.

Marder, Michael. 2013. *Plant-Thinking: A Philosophy of Vegetal Life*. New York: Columbia University Press.

Marx, Karl. 1959. "Estranged Labour." In *Economic and Philosophical Manuscripts of 1844*, trans. Martin Mulligan. Moscow: Progress Publishers. https://www.marxists.org/archive/marx/works/1844/manuscripts/labour.htm. Accessed July 8, 2021.

——. 1973. *Grundrisse*, trans. Martin Nicolaus. Harmondsworth, UK: Penguin in association with New Left Review. https://www.marxists.org/archive/marx/works/1857/grundrisse/ch13.htm. Accessed July 8, 2021.

Massumi, Brian. 2014. *What Animals Teach Us About Politics*. Durham, NC: Duke University Press.

Maturana, Humberto R., and Francisco J. Varela. 1980. *Autopoiesis and Cognition: The Realization of the Living*. Dordrecht, Netherlands: D. Reidel.

McGee, Kyle. 2014. *Bruno Latour: The Normativity of Networks*. Abingdon, UK: Routledge.

McGuire, K. L. 2007. "Common Ectomycorrhizal Networks May Maintain Monodominance in a Tropical Rain Forest." *Ecology* 88, no. 3: 567–74.

McIntyre, Niamh, and Mark Rice-Oxley. 2018. "It's 34,361 and Rising: How the List Tallies Europe's Migrant Bodycount." *Guardian*. June 30, 2018. https://www.theguardian.com/world/2018/jun/20/the-list-europe-migrant-bodycount.

McNay, Lois. 2016. "Agency." In *The Oxford Handbook of Feminist Theory*, ed. Lisa Disch and Mary Hawkesworth, 39–57. Oxford: Oxford University Press.

Mitchell, Timothy. 2013. *Carbon Democracy: Political Power in the Age of Oil*. London: Verso.

Modelling Kivalina (Andrea Bagnato, Helene Kazan, Hannah Meszaros Martin, Daniel Fernández Pascual, and Alon Schwabe). 2014. "The Coming Storm and the Changing Shoreline of Kivalina." In *Forensis: The Architecture of Public Truth*, Forensic Architecture, 700–11. Berlin: Sternberg Press & Forensic Architecture.

Moore, Jason W. 2015. *Capitalism in the Web of Life: Ecology and the Accumulation of Capital.* London: Verso.

Morozov, Evgeny. 2019. "Digital Socialism? The Calculation Debate in the Age of Big Data." *New Left Review* 116–117: 33–67.

Moten, Fred. 2015. "The Touring Machine (Flesh Thought Inside Out)." In *Plastic Materialities: Politics, Legality, and Metamorphosis in the Work of Catherine Malabou*, ed. Brenna Bhandar and Jonathan Goldberg-Hiller, 265–86 Durham, NC: Duke University Press.

Mouffe, Chantal. 2013. *Agonistics: Thinking the World Politically*. London: Verso.

Nagel, Thomas. 1974. "What Is It Like to Be a Bat?" *Philosophical Review* 83, no. 4: 435–50.

Nietzsche, Friedrich. 1987. *Thus Spoke Zarathustra*, trans. R. J. Hollingdale. London: Penguin.

——. 1998. *Twilight of the Idols*, trans. Duncan Large. Oxford: Oxford University Press.

Nisbet, Nancy D. 2018. "Event, Plasticity, and Mutation: Harnessing the Work of Malabou and Badiou in Support of a Molecular Event." In *Thinking Catherine Malabou: Passionate Detachments*, ed. Thomas Wormald and Isabell Dahms. London: Rowman & Littlefield. Online edition.

Nony, Anaïs. 2015. "Nootechnics of the Digital." *Parallax* 23, no. 2: 129–46.

Noys, Benjamin. 2010. *The Persistence of the Negative: A Critique of Contemporary Continental Theory*. Edinburgh: Edinburgh University Press.

O'Brien, Lilian. 2015. *Philosophy of Action*. Basingstoke, UK: Palgrave Macmillan.

Panagia, Davide. 2020. "On the Possibilities of a Political Theory of Algorithms." *Political Theory* 49, no. 1: 109–33.

Parisi, Luciana. 2019. *Contagious Architecture: For an Aesthetic Computation of Space*. Cambridge, MA: MIT Press.

Parikka, Jussi. 2010. *Insect Media: An Archaeology of Animals and Technology*. Minneapolis: University of Minnesota Press.

Passoth, Jan-Hendrik, Birgit Peuker, and Michael Schillmeier. 2012. "Introduction." In *Agency Without Actors? New Approaches to Collective Action*, ed. Jan-Hendrik Passoth, Birgit Peuker, and Michael Schillmeier, 1–11. Abingdon, UK: Routledge.

Pickering, Andrew. 2010. *The Cybernetic Brain: Sketches of Another Future*. Chicago: Chicago University Press.

Proctor, Hannah. 2011. "Neuronal Ideologies: Catherine Malabou's Explosive Plasticity in Light of the Marxist Psychology of A. R. Luria." *Dandelion* 2, no. 1: 1–13.

Rammert, Werner. 2012. "Distributed Agency and Advanced Technology. Or: How to Analyze Constellations of Collective Inter-Agency." In *Agency Without Actors? New Approaches to Collective Action*, ed. Jan-Hendrik Passoth, Birgit Peuker, and Michael Schillmeier, 89–112. Abingdon, UK: Routledge.

Rancière, Jacques. 2004. *Politics and Aesthetics: The Distribution of the Sensible*, ed. and trans. Gabriel Rockhill. London: Continuum.

——. 2009. *Dissensus: On Politics and Aesthetics*, ed. and trans. Steven Corcoran. London: Continuum.

Rescher, Nicholas. 1996. *Process Metaphysics: An Introduction to Process Philosophy*. Albany, NY: State University of New York Press.

Rid, Thomas. 2016. *The Rise of the Machines: The Lost History of Cybernetics*. Melbourne: Scribe.

Roberts, Ben. 2005. "Stiegler Reading Derrida: The Prosthesis of Deconstruction in Technics." *Postmodern Culture: An Electronic Journal of Interdisciplinary Criticism* 16, no. 1. https://muse.jhu.edu/article/192267.

Ross, Daniel. 2018. "Introduction." In *Bernard Stiegler, the Neganthropocene*, trans. Daniel Ross, 7–32. London: Open Humanities Press.

Rossiter, Ned, and Soenke Zehle. 2013. "Acts of Translation: Organized Networks as Algorithmic Technologies of the Common." In *Digital Labor: The Internet as Playground and Factory*, ed. Trebor Scholz, 225–39. New York: Routledge.

Rouvroy, Antoinette, and Thomas Berns. 2013. "Algorithmic Governmentality and Prospects of Emancipation." *Réseaux*, 177, no. 1: 163–96. https://www.cairn-int.info/article-E_RES_177_0163--algorithmic-governmentality-and-prospect.htm.

Scholz, Trebor. 2017. *Uberworked and Underpaid: How Workers Are Disrupting the Digital Economy*. Cambridge: Polity.

Schüll, Natasha Dow. 2016. "Data for Life: Wearable Technology and the Design of Self-Care." *BioSocieties* 11, no. 3: 317–33.

Sen, Amartya. 1997. *On Economic Inequality*. Oxford: Clarendon.

Shadlen, M.N., and A.L. Roskies. 2012. "The Neurobiology of Decision-Making and Responsibility: Reconciling Mechanism and Mindedness." *Frontiers in Neuroscience* 6, no. 56. https://doi.org/10.3389/fnins.2012.00056.

Simondon, Gilbert. 2013. *L'Individuation à lumière des notions de forme et d'information*. ("Individuation in the light of the notions of form and information.") Grenoble, France: Jérôme Millon.

Smith, Mick. 2013. "Ecological Community, the Sense of the World, and Senseless Extinction." *Environmental Humanities* 2: 21–41.

Sparrow, Tom. 2014. *Plastic Bodies: Rebuilding Sensation After Phenomenology*. London: Open Humanities Press.

Srinivasan, Amia. 2018. "What Termites Can Teach Us." *New Yorker*, September 10, 2018. https://www.newyorker.com/magazine/2018/09/17/what-termites-can-teach-us.

Srnicek, Nick, and Alex Williams. 2015. *Inventing the Future: Postcapitalism and a World Without Work*. London: Verso.

Stengers, Isabelle. 2011. *Cosmopolitics II*, trans. Robert Bononno. Minneapolis: University of Minnesota Press.

Stiegler, Bernard. 1998. *Technics and Time, 1: The Fault of Epimetheus*, trans. Richard Beardsworth and George Collins. Stanford: Stanford University Press.

——. 2003. "Technics of Decision: An Interview." *Angelaki* 8, no. 2: 151–68.

——. 2006. "*Chute et élévation: L'apolitique de* Simondon." ("Fall and rise: Simondon's apolitics.") *Revue philosophique de la France et de l'étranger* 131: 325–41.

——. 2009a. *Technics and Time, 2: Disorientation*, trans. Stephen Barker. Stanford: Stanford University Press.

——. 2009b. "The Theatre of Individuation: Phase-Shift and Resolution in Simondon and Heidegger." Trans. Kristina Lebedeva. *Parrhesia* 7: 46–57.

——. 2010a. *For a New Critique of Political Economy*, trans. Daniel Ross. Cambridge: Polity.

——. 2010b. *Taking Care of Youth and the Generations*, trans. Stephen Barker. Stanford: Stanford University Press.

——. 2011. *The Decadence of Industrial Democracies: Disbelief and Discredit, 1*, trans. Daniel Ross and Suzanne Arnold. Cambridge: Polity.

——. 2013a. *Uncontrollable Societies of Disaffected Individuals: Disbelief and Discredit, 2*, trans. Daniel Ross. Cambridge: Polity.

——. 2013b. *What Makes Life Worth Living: On Pharmacology*, trans. Daniel Ross. Cambridge: Polity.

——. 2014. *Symbolic Misery, 1: The Hyper-Industrial Epoch*, trans. Barnaby Norman. Cambridge: Polity.

——. 2015a. *Symbolic Misery, 2: The* 'katastrophè' *of the Sensible*, trans. Barnaby Norman. Cambridge: Polity.

——. 2015b. *States of Shock: Stupidity and Knowledge in the Twenty-First Century*, trans. Daniel Ross. Cambridge: Polity.

——. 2015c. "The Uncanniness of Thought and the Metaphysics of Penelop." Trans. Arne de Boever, with Greg Flanders and Alicia Harrison. *Parrhesia* 23: 63–77.

——. 2016. *Automatic Society, 1: The Future of Work*, trans. Daniel Ross. Cambridge: Polity.

——. 2017. "The New Conflict of the Faculties and Functions: Quasi-Causality and Singularity in the Anthropocene." *Qui Parle* 26, no. 1: 79–99.

——. 2018a. *The Neganthropocene*. London: Open Humanities Press.

——. 2018b. " 'We Have to Become the Quasi-Cause of Nothing—of *Nihil*': An Interview with Bernard Stiegler" (interview by Judith Wambacq, Daniel Ross, and Bart Buseyne). *Theory, Culture & Society* 35, no. 2: 137–56.

——. 2019. *The Age of Disruption: Technology and Madness in Computational Capitalism*, trans. Daniel Ross. Cambridge: Polity.

Stone, Christopher. 2010. *Should Trees Have Standing: Law, Morality and the Environment*. Oxford: Oxford University Press.

Sunstein, Cass. 2018. *#Republic: Divided Democracy in the Age of Social Media*. Princeton, NJ: Princeton University Press.

Szendy, Peter. 2013. *Kant in the Land of Extraterrestrials: Cosmopolitical Philosofictions*, trans. Will Bishop. New York: Fordham University Press.

Tavares, Paulo. 2014. "Nonhuman Rights." In *Forensis: The Architecture of Public Truth*, ed. Forensic Architecture, 533–72. Berlin: Sternberg Press & Forensic Architecture.

Terranova, Tiziana. 2013. "Free Labor." In *Digital Labor: The Internet as Playground and Factory*, ed. Trebor Scholz, 33–57. New York: Routledge.

——. 2014. "Red Stack Attack! Algorithms, Capital and the Automation of the Common." In *Accelerate: The Accelerationist Reader*, ed. Robin Mackay and Armen Avanessian, 379–98. Falmouth, UK: Urbanomic.

Tollefsen, Deborah. 2015. *Groups as Agents*. Cambridge: Polity.

Tufekci, Zeynep. 2017. *Twitter and Tear Gas: The Power and Fragility of Networked Protest*. New Haven, CT: Yale University Press.

Turner, Ben. 2016. "Life and the Technical Transformation of *Différance*: Stiegler and the Noopolitics of Becoming Non-Inhuman." *Derrida Today* 9, no. 2: 177–98.

Uexküll, Jakob von. 1957. "A Stroll Through the Worlds of Animals and Men: A Picture Book of Invisible Worlds." In *Instinctive Behavior: The Development of a Modern Concept*, trans. and ed. Claire H. Schiller, 5–80. New York: International Universities.

Usher, Phillip John. 2019. *Exterranean: Extraction in the Humanist Anthropocene*. New York: Fordham University Press.

Verbeek, Peter-Paul. 2005. *What Things Do: Philosophical Reflections on Technology, Agency, and Design*, trans. Robert P. Crease. University Park: Pennsylvania State University Press.

Vesco, Shawna. 2015. "Collective Disindividuation and/or Barbarism: Technics and Proletarianization." *Boundary 2* 42, no. 2: 85–104.

Wark, McKenzie. 2004. *A Hacker Manifesto*. Cambridge, MA: Harvard University Press.

——. n.d. "Digital Labor and the Anthropocene." *DIS Magazine*. http://dismagazine.com/disillusioned/discussion-disillusioned/70983/mckenzie-wark-digital-labor-and-the-anthropocene/. Accessed July 8, 2021.

Watkin, Christopher. 2017. *French Philosophy Today: New Figures of the Human in Badiou, Meillassoux, Malabou, Serres, and Latour*. Edinburgh: Edinburgh University Press.

——. 2020. *Michel Serres: Figures of Thought*. Edinburgh: Edinburgh University Press.

Watson, Matthew C. 2011. "Cosmopolitics and the Subaltern: Problematizing Latour's Idea of the Commons." *Theory, Culture & Society* 28, no. 3: 55–79.

Weizman, Eyal. 2012. *Forensic Architecture: Notes from Fields and Forums*. Ostfildern, Germany: Hatje Catz.

——. 2016. Response to "A Questionnaire on Materialisms." *October* 155: 104.

——. 2017. *Forensic Architecture: Violence at the Threshold of Detectability*. New York: Zone.

Whitehead, Alfred North. 1978. *Process and Reality*. New York: The Free Press.

Wilcox, Lauren. 2017. "Embodying Algorithmic War: Gender, Race, and the Posthuman in Drone Warfare." *Security Dialogue* 48, no. 1: 11–28.

Williams, James. 2018. *Stand Out of Our Light: Freedom and Resistance in the Attention Economy*. Cambridge: Cambridge University Press.

Williams, Raymond. 1977. *Marxism and Literature*. Oxford: Oxford University Press.

Wills, Tom. 2019. "Sweden: Rogue Algorithm Stops Welfare Payments for up to 70,000 Unemployed." Algorithm Watch, February 25, 2019. https://algo-rithmwatch.org/en/rogue-algorithm-in-sweden-stops-welfare-payments/.

INDEX

INSURRECTIONS: CRITICAL STUDIES IN RELIGION, POLITICS, AND CULTURE

SLAVOJ ŽIŽEK, CLAYTON CROCKETT, CRESTON DAVIS, JEFFREY W. ROBBINS, EDITORS

After the Death of God, John D. Caputo and Gianni Vattimo, edited by Jeffrey W. Robbins

The Politics of Postsecular Religion: Mourning Secular Futures, Ananda Abeysekara

Nietzsche and Levinas: "After the Death of a Certain God," edited by Jill Stauffer and Bettina Bergo

Strange Wonder: The Closure of Metaphysics and the Opening of Awe, Mary-Jane Rubenstein

Religion and the Specter of the West: Sikhism, India, Postcoloniality, and the Politics of Translation, Arvind Mandair

Plasticity at the Dusk of Writing: Dialectic, Destruction, Deconstruction, Catherine Malabou

Anatheism: Returning to God After God, Richard Kearney

Rage and Time: A Psychopolitical Investigation, Peter Sloterdijk

Radical Political Theology: Religion and Politics After Liberalism, Clayton Crockett

Radical Democracy and Political Theology, Jeffrey W. Robbins

Hegel and the Infinite: Religion, Politics, and Dialectic, edited by Slavoj Žižek, Clayton Crockett, and Creston Davis

What Does a Jew Want? On Binationalism and Other Specters, Udi Aloni

A Radical Philosophy of Saint Paul, Stanislas Breton, edited by Ward Blanton, translated by Joseph N. Ballan

Hermeneutic Communism: From Heidegger to Marx, Gianni Vattimo and Santiago Zabala

Deleuze Beyond Badiou: Ontology, Multiplicity, and Event, Clayton Crockett

Self and Emotional Life: Philosophy, Psychoanalysis, and Neuroscience, Adrian Johnston and Catherine Malabou

The Incident at Antioch: A Tragedy in Three Acts / L'Incident d'Antioche: Tragédie en trois actes, Alain Badiou, translated by Susan Spitzer

Philosophical Temperaments: From Plato to Foucault, Peter Sloterdijk

To Carl Schmitt: Letters and Reflections, Jacob Taubes, translated by Keith Tribe

Encountering Religion: Responsibility and Criticism After Secularism, Tyler Roberts

Spinoza for Our Time: Politics and Postmodernity, Antonio Negri, translated by William McCuaig

Force of God: Political Theology and the Crisis of Liberal Democracy, Carl A. Raschke

Factory of Strategy: Thirty-Three Lessons on Lenin, Antonio Negri, translated by Arianna Bove

Cut of the Real: Subjectivity in Poststructuralism Philosophy, Katerina Kolozova

A Materialism for the Masses: Saint Paul and the Philosophy of Undying Life, Ward Blanton

Our Broad Present: Time and Contemporary Culture, Hans Ulrich Gumbrecht

Wrestling with the Angel: Experiments in Symbolic Life, Tracy McNulty

Cloud of the Impossible: Negative Theology and Planetary Entanglements, Catherine Keller

What Does Europe Want? The Union and Its Discontents, Slavoj Žižek and Srećko Horvat

Harmattan: A Philosophical Fiction, Michael Jackson

Nietzsche Versus Paul, Abed Azzam

Christo-Fiction: The Ruins of Athens and Jerusalem, François Laruelle

Paul's Summons to Messianic Life: Political Theology and the Coming Awakening, L. L. Welborn

Reimagining the Sacred: Richard Kearney Debates God with James Wood, Catherine Keller, Charles Taylor, Julia Kristeva, Gianni Vattimo, Simon Critchley, Jean-Luc Marion, John Caputo, David Tracy, Jens Zimmermann, and Merold Westphal, edited by Richard Kearney and Jens Zimmermann

A Hedonist Manifesto: The Power to Exist, Michel Onfray

An Insurrectionist Manifesto: Four New Gospels for a Radical Politics, Ward Blanton, Clayton Crockett, Jeffrey W. Robbins, and Noëlle Vahanian

The Intimate Universal: The Hidden Porosity Among Religion, Art, Philosophy, and Politics, William Desmond

Heidegger: His Life and His Philosophy, Alain Badiou and Barbara Cassin, translated by Susan Spitzer

The Work of Art: Rethinking the Elementary Forms of Religious Life, Michael Jackson

Sociophobia: Political Change in the Digital Utopia, César Rendueles, translated by Heather Cleary

There's No Such Thing as a Sexual Relationship: Two Lessons on Lacan, Alain Badiou and Barbara Cassin, translated by Susan Spitzer and Kenneth Reinhard

Unbearable Life: A Genealogy of Political Erasure, Arthur Bradley